David Hilbert

Natur und mathematisches Erkennen

Vorlesungen, gehalten 1919–1920 in Göttingen

Nach der Ausarbeitung von Paul Bernays

Herausgegeben von David E. Rowe

Springer Basel AG

Adresse des Herausgebers

Professor Dr. David E. Rowe
Department of Mathematics
Pace University
Pleasantville, NY 10570
USA

Herausgeber und Verlag danken dem Mathematischen Institut Göttingen für die freundliche Erlaubnis zum Abdruck dieser Vorlesungen.

Einbandabbildung: Fotografie von David Hilbert, mit freundlicher Genehmigung des Springer-Verlag Archivs, New York.

Die Deutsche Bibliothek – CIP-Einheitsaufnahme

Hilbert, David:
Natur und mathematisches Erkennen: Vorlesungen, gehalten 1919–1920 in Göttingen – David Hilbert. Nach der Ausarbeitung von Paul Bernays. Hrsg. von David E. Rowe. – Basel ; Boston ; Berlin : Birkhäuser, 1992
ISBN 978-3-7643-2668-5 ISBN 978-3-0348-8640-6 (eBook)
DOI 10.1007/978-3-0348-8640-6
NE: Bernays, Paul [Bearb.]

Ursprünglich erschienen bei Birkhäuser Verlag , Basel 1992

Printed from the editor's camera-ready manuscript on acid-free paper in Germany

ISBN 978-3-7643-2668-5

Inhalt

INTRODUCTION

At the time of David Hilbert's death in 1943, his leading disciple, Hermann Weyl, wrote that "...the era of mathematics upon which he impressed the seal of his spirit and which is now sinking below the horizon achieved a more perfect balance than prevailed before and after, between the mastering of concrete problems and the formation of general abstract concepts."[1] Weyl attributed this "happy equilibrium" in no small part to Hilbert's work and its influence, adding that "no mathematician of equal stature has risen from our generation."[2]

Surely, it would be difficult to exaggerate the importance of Hilbert's contributions to twentieth-century mathematics or even to conceive of what mathematics today would be like without them. He overturned the conceptual framework of older fields ranging from invariant theory and algebraic number theory to the foundations of geometry. He rehabilitated the Dirichlet Principle, propelled integral equation theory to the forefront of active research, derived the field equations governing Einstein's general theory of relativity, created modern proof theory and metamathematics, and throughout his career he championed the power and efficacy of the axiomatic method not only for mathematics but for all of the exact sciences. Every educated mathematician knows something about Hilbert space, the Hilbert problems, and Hilbert's formalist program. If the mathematician happens to be an algebraist, then he can also give a fairly precise explanation of some version of Hilbert's finite basis theorems for algebraic invariants. If a number theorist, then he will certainly be aware that many key concepts in modern algebraic number theory stem from Hilbert's *Zahlbericht*. An algebraic geometer will not only know the statement but also a proof of Hilbert's *Nullstellensatz*. And a functional analyst might well go on for hours talking about important properties of Hilbert space.

Hilbert was not a particularly prolific mathematician. In fact, his output appears almost meager when placed alongside the collected works of someone like Poincaré, the only mathematician in his class around the turn of the century. Of course, at the pinnacle of mathematical achievement quantity counts far less than quality (as witnessed by the work of Abel, Galois, and Riemann), and on a qualitative scale Hilbert's work ranks very high indeed. Still, even taken together, the sum of all the impressive theories and results connected with his name fails to provide an adequate measure of the full scope, power, and influence of his mathematical career. Any such assessment must go beyond the pages of his published work and confront the man

---

[1] Weyl 1944, p. 612.

[2] *Ibid.*

himself; for in Hilbert's case, one truly cannot separate the messenger from his message.[3]

At the core of Hilbert's mathematical activity—from his appointment as a Königsberg *Privatdozent* in 1886 until his retirement from the Göttingen faculty in 1930—stood his lecture courses. These covered an immense range of material, nearly all of it directly related to his own research interests. Few mathematicians, if any, ever attained such a rich and harmonious balance between teaching and research as did Hilbert during his first two decades in Göttingen. There he stood in the middle of a stimulating scientific subculture, surrounded by a talented array of colleagues (including Klein, Minkowski, Runge, and Schwarzschild), admiring students, and a handful of intimate protégés. Between 1898 and 1914, Hilbert directed the doctoral dissertations of no fewer than sixty students, a staggering number even today; for the time, simply an incredible figure. Among the many distinguished names that adorn this long list, one finds: Otto Blumenthal, Max Dehn, Felix Bernstein, Erhard Schmidt, Ernst Hellinger, Hermann Weyl, Alfred Haar, Richard Courant, and Erich Hecke.[4] Still, Hilbert's influence extended far beyond the inner-circle of his immediate students. Through his lectures and published work he succeeded in galvanizing and directing the research interests of literally hundreds of mathematicians from all over the world.

Hilbert's strong-willed, resilient, and ever-optimistic personality was never more in evidence than when he delivered a course of lectures on "Natur und mathematisches Erkennen" in the autumn of 1919. These lectures, published for the first time in this volume, represent the clearest and most comprehensive statement of his mature views on the nature of mathematical and physical knowledge. The present edition is based on a typescript copy of Hilbert's lectures prepared by Hilbert's long-time assistant, Paul Bernays.[5] The Bernays' *Ausarbeitung*, located in the Library of the Mathematisches Institut in Göttingen, contains several handwritten emendations and comments, some of which appear in Hilbert's own hand. This latter circumstance lends additional authenticity to the Bernays' work, since it suggests

[3] On this point, see Constance Reid *Hilbert*, (New York: Springer, 1970).

[4] For the full list, see *Hilbert: GA*, 3: pp. 431-433.

[5] The present edition appears with the kind permission of the Mathematisches Institut der Universität Göttingen, which in 1990 prepared a new version of the Bernays' *Ausarbeitung* and made a limited number of copies of these available. My work on this project was partially supported by National Science Foundation Grant DIR 8821421. I would like to take this opportunity to thank Bernie Freydberg, Carolyn Geisel, Susann Hensel, John McCleary, and Abe Shenitzer for much appreciated technical assistance that helped make this volume possible.

that Hilbert had studied the manuscript and that he had found its contents accurately reflected the ideas and opinions he wished to express.

As can be seen from the Table of Contents, the conceptual framework Hilbert adopted for his lectures on "Natur und mathematisches Erkennen" follows a triadic pattern. Thus, in Parts I and II he pursued parallel critiques of conventional views about the nature of mathematical knowledge, on the one hand, and physical knowledge, on the other, in accordance with the tripartite conceptual schema of "Begriff, Urteil, Schluß." This set the stage for Part III, where he undertook a lengthy mathematical analysis of three interrelated problems—causality, probability, and ideality—explored in a manner designed to shed light on certain fundamental philosophical concerns. Taken together, these lectures constitute a sustained, tightly-interwoven, and largely self-contained argument, and they should be read as such. Nevertheless, to appreciate the full significance of Hilbert's reasoning, one must take into account the broader historical context that motivated these lectures. The remarks that follow are intended as hints in this direction.

—1—

If one picks up any standard text dealing with the foundations of mathematics, one learns that shortly after the turn of the century three principal schools of thought emerged to dominate philosophical deliberations about mathematics during the subsequent era: logicism, intuitionism, and formalism. Further along (perhaps the same page) one will find some statement identifying Hilbert as the founding father of the formalist school. Since most readers of this volume will already be aware of this, they may waive this exercise and move on to the next, namely, to read carefully what Hilbert says about the nature of mathematical knowledge in the first Chapter of "Natur und mathematisches Erkennen." For the views he expresses therein make it abundantly clear that his fundamental conception of mathematics was diametrically opposed to the formalist position that logicians and philosophers of mathematics have so often associated with his name. In fact, this opening section is devoted to one long, devastating critique of the notion that mathematics can be *understood* as or *reduced* to a game played with a system of formal rules for manipulating a collection of meaningless signs. Not that this should come as any great surprise to those familiar with the larger context of Hilbert's work. As Egbert Brieskorn once remarked, "ein so großer Mathematiker wie Hilbert konnte einfach nicht die Mathematik als bedeutungsloses Spiel ansehen...."[6]

[6] He went on to say: "Leider haben geringere Geister mit dem Namen 'Formalismus' dann ihr ganz anderes Verständnis von der Mathematik als Formelspiel propagiert..." (Brieskorn, p. 228.)

According to James Henle, the formalist doctrine can be expressed as a single metatheorem which asserts that "...all of pure mathematics can be imbedded in formal systems."[7] Does this definition make Hilbert a formalist? Perhaps, but the salient point is that such a characterization fails to describe his larger views or to tell us anything of significance about his "real" philosophy of mathematics, which was clearly grounded in deeper questions concerning mathematics as a higher form of human knowledge. Although Hilbert apparently did not have a particularly detailed knowledge of the writings of leading philosophers, one can easily discern the influence of Kant's epistemology on his thought, as, for example, in his argument (see Chapters 7 and 8 of "Natur und mathematisches Erkennen") that causality is conditioned by human perceptions and purposes. Moreover, recent investigations of Volker Peckhaus have fully documented Hilbert's long-standing interest in the work of the philosopher, Leonard Nelson, a leading representative of the philosophical tradition established by Jacob Friedrich Fries.[8]

If philosophers of mathematics have been too busy packaging and promoting their favorite brands of formalism to notice that Hilbert's views bear little resemblance with their own, historians have not fared much better. In his recent, thought-provoking study, ***Moderne—Sprache—Mathematik***, Herbert Mehrtens offers a sweeping panoramic analysis of philosophical, social, and ideological issues that shaped the debates on foundations during the Weimar era. However, his characterization of Hilbert's position—cryptically expressed by the dictum "die Mathematik ist Arbeit an der Sprache"[9]—diverges markedly from the views expressed in "Natur und mathematisches Erkennen."

At the very outset of these lectures, Hilbert describes his goal as "eine Art Vorbereitung für eine Erkenntnistheorie." The emphasis, however, falls on scrutinizing the very latest advances in both mathematics and physics in an effort to shed light on fundamental epsitemological problems. When considering the nature of mathematical knowledge, Hilbert tends, for the most part, to avoid essentialist arguments, preferring to illuminate the issue of what mathematics is by showing *how* it is done. The closest he comes to offering an essentialist definition of mathematics is in the following passage from Chapter 2:

> Das Wesentliche der mathematischen Methode besteht
> in der konsequenten Ausgestaltung des Verfahrens, das

[7] Henle, p. 13.

[8] See Peckhaus, pp. 123-224.

[9] Mehrtens, p. 135

> dem formalen Denken eigentümlich ist, also des logischen Denkens überhaupt. Dem Denken steht ein unermeßlicher Vorrat an formalen Beziehungen zur Verfügung, und es kommt darauf an, solche Systeme von formalen Beziehungen zu finden, die sich den in der Wirklichkeit vorgefundenen Beziehungen anpassen lassen.

Another striking passage occurs at the conclusion of a lengthy discussion in Chapter 1, in which Hilbert argues against employing a non-Archimedian number system as a model for continuous phenomena:

> Nun werden wir... diese [Archimedische] Forderung ohne zwingenden Anlaß nicht preisgeben. Und ein solcher Anlaß liegt in unserer Erfahrung auch gar nicht vor, vielmehr besteht die bemerkenswerte Tatsache, daß alle Abstände in der Natur miteinander zahlenmäßig vergleichbar sind. Die größten astronomischen Entfernungen sowie die kleinsten Abstände, wie wir sie bei den Atomen und Elektronen finden, lassen sich alle durch Vielfache und Bruchteile eines Zentimeters ausdrücken. Daher ist das Kontinuum dritter Art, welches als Zahlensystem in der Mathematik natürlich seine Bedeutung hat, für den Zweck der mathematischen Auffassung des Stetigen abzulehnen, und wir bleiben also bei dem Kontinuum zweiter Art stehen.
>
> So sehen wir an dem Beispiel des Kontinuums, wie die mathematische Begriffsbildung durch die Anschauung angeregt und von der Erfahrung geleitet wird. Es ließen sich hierfür noch zahlreiche Beispiele geben.

All of these remarks suggest that Hilbert's philosophy had more in common with the position of so-called quasi-empiricist thinkers than with the standard tenets of formalism.[10] In view of recent attacks on formalism and the growing interest in empirically-grounded philosophies of mathematics, a serious reassessment of Hilbert's thought and its relevance for current discussions would appear to be long overdue.[11]

At the heart of Hilbert's philosophical outlook stood his belief in the fundamental unity and harmony of mathematical ideas. He rejected the notion that mathematics can be subdivided into autonomous fields with no meaningful interplay between them. At the conclusion of his famous lecture,

---

[10] For a sampling of quasi-empiricist opinion, see the volume edited by Tymoczko.

[11] The recent studies of Toepell and Peckhaus represent significant contributions in this direction.

"Mathematische Probleme," delivered in 1900 at the Second International Congress of Mathematicians held in Paris, he confronted this particular issue head on, by asserting that:

> ... die mathematische Wissenschaft ist meiner Ansicht nach ein unteilbares Ganzes, ein Organismus, dessen Lebensfähigkeit durch den Zusammenhang seiner Teile bedingt wird. Denn bei aller Verschiedenheit des mathematischen Wissenstoffes im einzelnen gewahren wir doch sehr deutlich die Gleichheit der logischen Hilfsmittel, die Verwandtschaft der Ideenbildungen in der ganzen Mathematik und die zahlreichen Analogien in ihren verschiedenen Wissensgebieten. Auch bemerken wir: je weiter eine mathematische Theorie ausgebildet wird, desto harmonischer und einheitlicher gestaltet sich ihr Aufbau, und ungeahnte Beziehungen zwischen bisher getrennten Wissenszweigen werden entdeckt. So kommt es, daß mit der Ausdehnung der Mathematik ihr einheitlicher Charakter nicht verlorengeht, sondern desto deutlicher offenbar wird.[12]

Hilbert went even further than this, suggesting that, unlike other disciplines, the inner unity of mathematical knowledge would enable future researchers to master its many branches. Thus, despite the impending explosion of new results and theories, he remained optimistic:

> ... Ich möchte... darauf hinweisen, wie sehr es im Wesen der mathematischen Wissenschaft liegt, daß jeder wirkliche Fortschritt stets Hand in Hand geht mit der Auffindung schärferer Hilfsmittel und einfacherer Methoden, die zugleich das Verständnis früherer Theorien erleichtern und umständliche ältere Entwicklungen beseitigen, und daß es daher dem einzelnen Forscher, indem er sich diese schärferen Hilfsmittel und einfacheren Methoden zu eigen macht, leichter gelingt, sich in den verschiedenen Wissenszweigen der Mathematik zu orientieren, als dies für irgendeine andere Wissenschaft der Fall ist.
>
> Der einheitliche Charakter der Mathematik liegt im inneren Wesen dieser Wissenschaft begründet; denn die Mathematik ist die Grundlage alles exakten naturwissenschaftlichen Erkennens.[13]

---

[12] Hilbert 1900b, p. 329.

[13] *Ibid.*

Hilbert's philosophy of mathematics reflected not only this universalism but also his faith in a preestablished harmony mediating between mathematical ideas and physical reality. His long-standing interest in the foundations of physics found expression in several largely-forgotten publications that followed on the heels of his work on integral equation theory. The first of these papers (Hilbert 1912, 1913, 1914) dealt with a consistency problem involving Kirchhoff's principle for the emission and absorption of radiant energy and its relationship to the wave theory of light. Immediately afterward, Hilbert turned his attention to the foundations of general relativity theory, a field in which he contributed significant new results.[14] Throughout this entire period, he covered a wide variety of topics from physics in his lecture courses in Göttingen, ranging from mechanics and electromagnetic theory to the structure of matter, kinetic gas theory, and the foundations of physics.[14]

It is important to keep in mind the extraordinary events that were transforming the world during this era. In 1919, the year Hilbert delivered his lectures on "Natur und mathematisches Erkennen," the European political order was entering a new as yet uncertain era. The terms of the Versailles Treaty, signed in the summer of 1919 and scheduled to go into effect on 1 January, 1920, evoked protests throughout Germany, as factions on the left and right vied to fill the power vacuum created by the abdication of the Kaiser. At the same time, a scientific revolution of profound significance created a fresh new chapter in man's conception of the universe.

On 6 November, 1919, the results of two crucial tests of Einstein's theory of general relativity were officially presented at a joint meeting of the Royal Society and the Royal Astronomical Society in London.[15] The test results came from two scientific expeditions from which photographs were taken of a lunar eclipse that occurred on 29 May, 1919. The data confirmed Einstein's prediction that a light ray passing through the vicinity of the sun's gravitational field would bend by a tiny amount (approximately 1.74 seconds of arc). Alfred North Whitehead, who attended the London meeting in which these findings were officially announced, gave this vivid eye-witness account of what transpired:

> The whole atmosphere of tense interest was exactly that of the Greek drama. We were the chorus commenting

[14] See Hilbert 1915, 1917 and the discussion in J. Mehra's study.

[14] Regarding these lectures, Otto Blumenthal once remarked: "Es wäre sehr zu wünschen, daß Teile dieser Vorlesungen veröffentlicht würden. Sie sollen nicht physikalische Pionierarbeit leisten, aber für die Sicherung des von den Physikern Gewonnenen und die Ordnung und das Verständnis der darin enthaltenen Beziehungen ist die erstrebte Säuberung wesentlich." Blumenthal, p. 417.

[15] See A. Pais, p. 305.

> on the decree of destiny as disclosed in the development of a supreme incident. There was dramatic quality in the very staging—the traditional ceremonial, and in the background the picture of Newton to remind us that the greatest of scientific generalizations was now, after more than two centuries, to receive its first modification. Nor was the personal interest wanting; a great adventure in thought had at length come home to shore.[16]

Interestingly enough, Hilbert only mentions this dramatic confirmation of Einstein's theory once, in Chapter 6 of " Natur und mathematisches Erkennen," and his description of the experiment and its outcome could hardly have been more restrained. This passage is all the more striking when contrasted with his earlier remarks at the end of Chapter 4, where he calls the principle of general relativity the "höchste Triumph des Geistes über die Erscheinungswelt," and at the close of Chapter 5, when he refers to this principle as the single best example of an eternal natural law. These circumstances all clearly suggest that Hilbert found Einstein's theory so theoretically compelling that he could not imagine abandoning it even if there had been no strong experimental evidence backing it up. His attitude must have been similar to Einstein's own, for when asked what his reaction would have been had the expeditions failed to confirm his theory, the latter reportedly retorted, "Da könnt' mir halt der liebe Gott leid tun, die Theorie stimmt doch."[17]

—2—

The rich blend of mathematical and physical ideas that runs like a thread through Hilbert's "Natur und mathematisches Erkennen," while one of the main qualities that make these lectures unique, can, of course, be found in his published work as well. In particular, his lecture on "Axiomatisches Denken" (Hilbert 1918), delivered in Zürich on 11 September, 1917 before a meeting of the Swiss Mathemematical Society, came closest to capturing the thematic flavor of these more extensive lectures. Here one encounters the same kind of conceptual analysis of mathematics and physics stressing their mutual interdependence and the role of axiomatization in both realms. Hilbert starts with the assertion that a mathematical or physical theory constitutes a system of ideas possessing a certain structure ("Fachwerk von Begriffen"). As the theory matures, certain key ideas emerge, serving as foundational principles from which the remaining results can be derived. Thus, for example, elementary number theory requires only the rules of arithmetic for

[16] Whitehead, p. 13.

[17] Bernstein, p. 144.

the natural numbers. The theory of prime numbers, on the other hand, depends on a deep theorem regarding the zeros of the Riemann ζ-function. The theory of statics requires only the parallelogram law for the composition of forces, whereas mechanics derives from the Lagrangian differential equations of motion, and electrodynamics is based on the Maxwell equations. Hilbert mentions many more cases, but emphasizes that, in each and every one, these key organizing principles gave way to even more fundamental ideas once the main results of the theory had been established. Thus, the original axioms, which arose spontaneously, as it were, in the course of the theory's development, proved to be provisional in nature, in so far as first principles were concerned. Only during the next stage, when researchers attempted to provide deeper foundations for the theory, did the process of axiomatization actually begin. Hilbert described the manner in which this organic process had penetrated into scientific thought as follows:

> So entstanden die eigentlichen heute sogenannten *Axiome* der Geometrie, der Arithmetik, der Statik, der Mechanik, der Strahlungstheorie oder der Thermodynamik. Diese Axiome bilden eine tiefer liegende Schicht von Axiomen gegenüber derjenigen Axiomschicht, wie sie durch die vorhin genannten zuerst zugrunde gelegten Sätze in den einzelnen Wissensgebieten charakterisiert worden ist. Das Verfahren der axiomatischen Methode... kommt also einer *Tieferlegung der Fundamente* der einzelnen Wissensgebiete gleich, wie eine solche ja bei jedem Gebäude nötig wird in dem Maße, als man dasselbe ausbaut, höher führt und dennoch für seine Sicherheit bürgen will.[18]

This architectural metaphor had an evident appeal to Hilbert; one finds it again, for instance, in the closing paragraph of Chapter 2. In his unpublished 1905 lectures, "Logische Principien des mathematischen Denkens," he employed a somewhat different version of the metaphor in connection with the paradoxes of set theory. This earlier rendering illuminates the broader context in which Hilbert viewed the process of axiomatization:

> Es ist in der Entwicklungsgeschichte der Wissenschaft wohl immer so gewesen, dass man ohne viele Scrupel eine Disciplin zu bearbeiten begann und soweit vordrang wie möglich, dass man dabei aber, oft erst nach langer Zeit, auf Schwierigkeiten stieß, durch die man gezwungen wurde, umzukehren und sich auf die Grundlagen der Disciplin zu besinnen. Das Gebäude der Wissenschaft wird nicht aufgerichtet wie ein Wohnhaus, wo

[18] Hilbert 1918, p. 148.

> zuerst die Grundmauern fest fundiert werden und man dann erst zum Auf- und Ausbau der Wohnräume schreitet; die Wissenschaft zieht es vor, sich möglichst schnell wohnliche Räume zu verschaffen, in denen sie schalten kann, und erst nachträglich, wenn es sich zeigt, dass hier und da die locker gefügten Fundamente den Ausbau der Wohnräume nicht zu tragen vermögen, geht sie daran, dieselben zu stützen und zu befestigen. Das ist kein Mangel, sondern die richtige und gesunde Entwicklung.[19]

In "Axiomatisches Denken," Hilbert went on to discuss two prime considerations for the logical foundations of any scientific theory. First, he prescribed an analysis of the logical interrelations between the various propositions that comprise a theory, i. e. their mutual dependence or independence. Secondly, he emphasized the desirability of establishing some guarantee of their mutual validity in the form of a consistency test. Both of these desiderata clearly have a strong bearing on the particular system of axioms that serve as the grounding principles for the theory in question. Within the physical realm, Hilbert pointed out an additional factor, namely the necessity of showing that the set of axioms adopted for one theory also be consistent for any related theory. In this connection, he mentioned his own proof of the consistency of the axioms for Kirchhoff's theory of emission and absorption of radiation with the fundamental laws of electromagnetic theory. As further examples of the same kind of consistency, he alluded to the kinetic theory of gases, which (as Boltzmann had shown) dovetails nicely with the fundamental laws of thermodynamics, and Einstein's theory of gravitation, which can in one sense be viewed as consistent with the classical Newtonian theory since it subsumes the latter as a limiting case. Quantum theory, on the other hand, had led to new laws inconsistent with classical electrodynamical theory as developed from Maxwell's equations, thus necessitating a revision of the axioms for electrodynamics.

Of course, the parallelism Hilbert expounded between the evolution of mathematical versus physical theories could only be carried out to a point. Thus, the symmetry he tried to maintain throughout the first two parts of "Natur und mathematisches Erkennen" begins to look rather strained by the time he reaches Chapter 6, which deals with the manifold relationships between theory and experiment in physics. The latter clearly have little in common with the purely self-referential arguments employed in mathematics. This distinction, however obvious, holds particular significance for the

---

[19] Peckhaus, p. 51.

question of inner-consistency. Because mathematical theories lack an empirical component, when contradictions (in the form of paradoxes) arise within them, the status of the theory as a whole can easily be called into question.

Hilbert tended to minimize the danger posed by the discovery of contradictions within a mathematical theory, characterizing these difficulties as the inevitable growing pains associated with a new field. However, the paradoxes of set theory forced him to take a hard look at the nature of mathematical arguments, the subject of his 1905 lectures on "Logische Principien des mathematischen Denkens." Even though his proof theory remained but a torso until the early 1920's, Hilbert's optimism never appeared to waver. Indeed, the "Wohnhaus" metaphor succinctly expressed his fundamental faith throughout this whole period. This same orientation also motivated the "historicism" in Chapter 3 of "Natur und mathematisches Erkennen," where he attempted to draw an analogy between the conceptual difficulties that plagued eighteenth-century analysis and the more recent crisis that had arisen in set theory.

When Hilbert turned to discuss the most pressing issues confronting the foundations of mathematics in "Axiomatisches Denken," he highlighted two in particular: the consistency of the axioms for arithmetic and for the system of axioms Ernst Zermelo had established for set theory in 1908. He further emphasized that both of these problems were intimately wedded to a whole complex of deep and difficult epistemological questions of "spezifisch mathematischer Färbung," among which he mentioned:

> ...das Problem der prinzipiellen *Lösbarkeit einer jeden mathematischen Frage*, das Problem der nachträglichen *Kontrollierbarkeit* des Resultates einer mathematischen Untersuchung, ferner die Frage nach einem *Kriterium für die Einfachheit* von mathematischen Beweisen, die Frage nach dem Verhältnis zwischen *Inhaltlichkeit und Formalismus* in Mathematik und Logik und endlich das Problem der *Entscheidbarkeit* einer mathematischen Frage durch eine endliche Anzahl von Operationen.[20]

Noteworthy about this passage is that Hilbert clearly distinguished between the first-named problem and the last. This suggests that he had by now begun to refine his thinking about the so-called *Entscheidungsproblem*.[21] In the loosely-formulated terminology of "Mathematische Probleme," he had spoken of the "Axiom von der Lösbarkeit eines jeden Problems," by which he meant "... daß ihre Lösung durch eine endliche Anzahl rein logischer Schlüsse gelingen muß."[22] The *Entscheidungsproblem* held an especially important

---

[20] Hilbert 1918, p. 153.

[21] See Moore, p. 7.

[22] Hilbert 1900b p. 297.

place in Hilbert's philosophical thought, but he hoped and believed that future investigations into the axiomatic foundations of logic would eventually clarify and resolve all of these related problems. As he summed up his position at the conclusion of "Axiomatisches Denken":

> Alle solchen prinzipiellen Fragen... scheinen mir ein wichtiges, neu zu erschließendes Forschungsfeld zu bilden, und zur Eroberung dieses Feldes müssen wir—das ist meine Überzeugung—den Begriff des spezifisch mathematischen Beweises selbst zum Gegenstand einer Untersuchung machen, gerade wie auch der Astronom die Bewegung seines Standortes berücksichtigen, der Physiker sich um die Theorie seines Apparates kümmern muß und der Philosoph die Vernunft selbst kritisiert.[23]

While conceding that, for the present time, this program remained but a sketch for future research, Hilbert reiterated the fundamental convictions that bolstered his optimistic assessment:

> Ich glaube: Alles, was Gegenstand des wissenschaftlichen Denkens überhaupt sein kann, verfällt, sobald es zur Bildung einer Theorie reif ist, der axiomatischen Methode und damit mittelbar der Mathematik. Durch Vordringen zu immer tieferliegenden Schichten von Axiomen im vorhin dargelegten Sinne gewinnen wir auch in das Wesen des wissenschaftlichen Denkens selbst immer tiefere Einblicke und werden uns der Einheit unseres Wissens immer mehr bewußt. In dem Zeichen der axiomatischen Methode erscheint die Mathematik berufen zu einer führenden Rolle in der Wissenschaft überhaupt.[24]

These remarks clearly echoed those Hilbert sounded at the conclusion of his Paris address, quoted earlier. The sentiment he expressed on both occasions brings to mind a penetrating comment once made by Hermann Weyl: "Hilbert is the champion of axiomatics.... Sometimes when he praised the axiomatic method he seemed to imply that it was destined to obliterate completely the constructive or genetic method. I am sure that, at least in later life, this was not his true opinion."[25] The affirmation of this last statement can be found throughout the pages of "Natur und mathematisches Erkennen." To see this, it may be helpful to present a brief synopsis of the main highlights and principal arguments Hilbert gave in the concluding three chapters of these lectures.

---

[23] Hilbert 1918, p. 155.

[24] *Ibid.*, p. 156.

[25] Weyl 1944, p. 644.

—3—

After completing his critique in Parts 1 and 2 of conventional wisdom regarding the nature and role of mathematical and physical knowledge in human thought, Hilbert turned to give an assessment of the mathematical and physical underpinnings of the concept of causality. Thus, in Chapter 7, he begins with an argument that purports to show why any description of physical reality must involve the quest to find physical laws. With regard to the concept of causality, he presses for an additional uniformity property, namely, that the same conditions will always produce the same result. When he asserts that this property has nothing whatsoever to do with regularity ("Gesetzlichkeit"), he means, of course, in the strictly logical sense. He thus proceeds to show that neither of these two notions implies the other.

Next follows a discussion of what one means mathematically by a description of given states ("Umständen") in a physical process. These states are represented by a system of functions:

$$\begin{aligned} u_1 &= f_1(x,y,z,t) \\ \vdots &= \vdots \\ u_n &= f_n(x,y,z,t) \end{aligned}$$

in the space-time coordinates (x,y,z,t). The uniformity condition for the process then leads to consideration of a system of partial differential equations:

$$\begin{aligned} \frac{\partial u_1}{\partial t} &= F_1(u_1,\ldots,u_n,\frac{\partial u_1}{\partial x},\frac{\partial u_1}{\partial y},\ldots,\frac{\partial u_n}{\partial z},\ldots) \\ \vdots &\quad \vdots \\ \vdots &\quad \vdots \\ \frac{\partial u_n}{\partial t} &= F_n(u_1,\ldots,u_n,\frac{\partial u_1}{\partial x},\frac{\partial u_1}{\partial y},\ldots,\frac{\partial u_n}{\partial z},\ldots) \end{aligned}$$

Such a system possesses two key properties: first, that the laws it describes are independent of the space and time coordinates; and, second, that the changes in physical state are only affected by local spatial information and not by temporal conditions.

Having analyzed the mathematical significance of the uniformity condition, Hilbert turned to the second prerequisite for a causal structure: regularity. This meant, in effect, that the fundamental functions F_i had to be derivable from some kind of mathematical law. Hilbert noted, however, that this condition was far too weak. To be physically significant, he insisted that the F_i conform to two principles: 1) they should arise from a variational problem; and 2), they must satisfy the principle of general covariance. Although these stipulations could always be met formally, Hilbert maintained that a significant physical theory would lead to a simpler, more elegant mathematical formulation. These principles operated "unconsciously" in Maxwell's theory, but afterward Einstein applied them "ganz bewußt und planmässig."

Of course, the simplicity of the laws of nature implies nothing about the simplicity of their action, as witnessed by the 3-body problem. To obtain specific solutions, Hilbert noted, one must introduce various intitial-value or boundary-value conditions on the system of differential equations. In connection with Einstein's theory, he pointed out that even after specification of boundary-value conditions the solution functions still contain arbitrary parameters. This mathematical feature of Einstein's theory reflected a "grundsätzliche philosophische Bedeutung." For if one could, in fact, introduce constraints that led to a unique set of solutions to the system of partial differential equations, one would then have a completely deterministic world—what Hilbert calls an Hegelian physics.

At the conclusion of Chapter 7, Hilbert returns to the question of causality, noting that the concept presupposes a third condition, namely a direction in time, for, without this, one could not differentiate between cause and effect. The possibility of distinguishing past from future plays a central part in thermodynamics, but not in other branches of physics in which the fundamental laws remain valid when one replaces t by $-t$ in their governing equations. These considerations lead into Chapter 8 where Hilbert examines the role of probabilistic laws in physics.

Hilbert's primary focus in Chapter 8 concerns the kinetic theory of gases and the conceptual difficulties connected with so-called irreversible physical processes. After a prolonged discussion of several fundamental questions concerning the periodicity and regularity of natural phenomena, he turns to a consideration of the probabilistic underpinnings of our physical knowledge. The principle point of this analysis aims, again, to draw an epistemological conclusion, namely, that there are no truly irreversible processes in nature in the strict sense, and that our sense of a time direction derives from observing processes with very special initial– or boundary–value conditions. Hilbert further suggests that, beyond the kinetic theory of gases, this same principle appears to hold true for non-mechanical phenomena, and that only in the field of organic nature does a direction in time become truly manifest. Since irreversible events arise, in the final analysis, from probabilistic considerations, the laws of thermodynamics which explain them fail to meet Hilbert's strictures for a physical law capable of distinguishing a direction in time. He concludes from this that the notion of causality is merely a most convenient way for human beings to organize physical knowledge, but that it is by no means a necessary attribute of physical processes.

The conceptual advance afforded by this distinction can be seen from Hilbert's remarks about the anthropomorphic nature of causality arguments in physics. He contrasts this kind of knowledge with man's higher knowledge of certain transcendent attributes of nature, for example, the principle of relativity or the universal constants. In essence, the process he describes in the final pages of Chapter 8 concerns man's long quest not only to uncover the laws regulating natural phenomena but to frame an understanding of these laws that will no longer reflect the effects of the sensory and cognitive equipment that enable human beings to study these same phenomena. The journey that set out by viewing the world via the cruder forms of anthropomorphism—animism and astrology—and which went on to explore nature's deeper "meaning" in the form of least action principles or the teleological arguments so popular in the eighteenth century, had now arrived at a new station. From this fresh vantage point, the very notion of causality so fundamental to all human understanding of nature appeared inextricably bound to the circumstances that precondition all human experience.

In Chapter 9, the discussion turns to another fundamental feature underlying both mathematics and physics: the appeal to "ideal structures." In view of the significance this notion plays in Hilbert's subsequent work on proof theory, this chapter should be regarded as must reading for those interested in the original motivation behind his work on foundations. Hilbert gives two famous historical examples to illustrate the manner in which ideal elements have entered into mathematical theories. In projective geometry, one adjoins an ideal "line at infinity" whose "ideal points" are determined by systems of "intersecting" parallel lines; in number theory, Kummer introduced his notion of ideal numbers (later reformulated by Dedekind into modern ideal theory), thereby effectively launching the theory of algebraic number fields. These innovations both produced remarkable conceptual unification, reflected in the duality theorems of projective geometry and in the prime factorization theorem for algebraic number fields.

In connection with physics, Hilbert makes clear that *Idealisierung* arises in a different context, "als Annäherung an ein Ideal." Again, he addresses the theme of removing the various vestiges of "Anschauung" (as in the mechanistic view) and anthropomorphism from physics. Thus, he underscores the importance of Planck's effort to frame natural laws by employing the universal constants of nature rather than arbitrary constants dependent on human conventions.

In concluding, Hilbert marshals forth a number of his prior conclusions while mounting a strong attack on Weyl's newly proposed unified field theory,[26] which he calls "eine extreme Idealisierung." This interpretation appears most curious given that Hilbert's earlier work on the foundations of

[26] Weyl 1918.

general relativity theory had attempted to erect a similar unified field theory based on the theory of matter of Gustav Mie. This theory employed a generalized form of Hamilton's principle, a formulation thus based on variational analysis, whereas Weyl introduced a gauge structure to represent the electromagnetic field components. Apart from these differences, however, the two theories aimed to accomplish nearly the same end.[27] In "Natur und mathematisches Erkennen," Hilbert criticizes Weyl's new approach as, in effect, the latest incarnation of "Hegelian physics," and thus a futile reductionist exercise. In reiterating the arguments that can be made against all such theories on general philosophical grounds, Hilbert mentions a number of paradoxes that arise in this connection: the impossibility of distinguishing a direction in time, even for living creatures; the complete determinism of all events and all thoughts; and the reduction of mind to a mere epiphenomenon. These views may not appear to jibe well with the thrust of some of his earlier pronouncements, but they certainly reveal that Hilbert opposed reductionism in physics just as strongly as he did in mathematics. With his final remarks, emminently quotable in this regard, he closed the loop and went back to his central concern:

> Bemerkenswert ist aber, daß wir bei der Erörterung der Methode der Physik durch die vorgefundenen Paradoxien auf das allgemeine philosophische Problem geführt werden, ob und wie es möglich ist, unser Denken durch das Denken selbst zu begreifen und es von jeglichen Paradoxien zu befreien. Es ist dies dieselbe Frage, welche auch unseren Bemühungen im Gebiete der mathematischen Logik zugrunde liegt.

David E. Rowe
Pleasantville, New York
September, 1991

[27] See the discussion in Mainzer, pp. 395-403.

REFERENCES

Jeremy Bernstein, 1973. *Einstein*, New York: Viking.

Otto Blumenthal, 1935. "Lebensgeschichte," in *Hilbert: GA*, 3: 388-429.

Egbert Brieskorn, 1974. "Über die Dialektik in der Mathematik," in Michael Otte, Hrsg., *Mathematiker über die Mathematik*, Berlin: Springer, 221-286.

James M. Henle, 1991. "The Happy Formalist," *Mathematical Intelligencer*, 13(1): 12-18.

David Hilbert, 1899. *Grundlagen der Geometrie*, Leipzig: Teubner.

———— 1900a. "Über den Zahlbegriff," *Jahresbericht der Deutschen Mathematiker-Vereinigung*, 8: 180-184.

———— 1900b. "Mathematische Probleme. Vortrag, gehalten auf dem internationalen Mathematiker-Kongreß zu Paris 1900," *Nachrichten von der königlichen Gesellschaft der Wissenschaften zu Göttingen. Mathematisch-physikalische Klasse aus dem Jahre 1900*, 253-297; reprinted in *Hilbert: GA*, 3: 290-329.

———— 1904. "Über die Grundlagen der Logik und der Arithmetik," *Verhandlungen des III. Internationalen Mathematiker-Kongresses in Heidelberg 1904*, Leipzig: Teubner: 174-185.

———— 1912. "Begründung der elementaren Strahlungstheorie," *Nachrichten von der königlichen Gesellschaft der Wissenschaften zu Göttingen. Mathematisch-physikalische Klasse aus dem Jahre 1912*: 773-789; reprinted in *Hilbert: GA*, 3: 217-230.

———— 1913. "Bemerkungen zur Begründung der elementaren Strahlungstheorie," *Nachrichten von der königlichen Gesellschaft der Wissenschaften zu Göttingen. Mathematisch-physikalische Klasse aus dem Jahre 1913*: 409-416; reprinted in *Hilbert: GA*, 3: 231-237.

———— 1914. "Zur Begründung der elementaren Stahlungstheorie," *Nachrichten von der königlichen Gesellschaft der Wissenschaften zu Göttingen. Mathematisch-physikalische Klasse aus dem Jahre 1914*: 275-298; reprinted in *Hilbert: GA*, 3: 238-257.

———— 1915. "Die Grundlagen der Physik, (Erste Mitteilung)" *Nachrichten von der königlichen Gesellschaft der Wissenschaften zu Göttingen. Mathematisch-physikalische Klasse aus dem Jahre 1915*: 395-407.

———— 1917. "Die Grundlagen der Physik, (Zweite Mitteilung)" *Nachrichten von der königlichen Gesellschaft der Wissenschaften zu Göttingen. Mathematisch-physikalische Klasse aus dem Jahre 1917*: 53-76.

———— 1918. "Axiomatisches Denken," *Mathematische Annalen*, 79: 405-415; reprinted in *Hilbert: GA*, 3: 146-156.

——— 1922. "Neubegründung der Mathematik. Erste Mitteilung," *Abhandlungen aus dem Mathematischen Seminar der Hamburgischen Universität*, 1: 157-177; reprinted in *Hilbert: GA*, 3: 157-177.

——— 1925. "Über das Unendliche," *Mathematische Annalen*, 95, 161-190.

——— 1932, 1933, 1935. *Gesammelte Abhandlungen* (*Hilbert: GA*), 3 vols. Berlin: Springer.

——— 1971. *Foundations of Geometry*, Leo Unger, trans. from 10th German ed., LaSalle, Ill.: Open Court.

David Hilbert and Wilhelm Ackermann, 1928. *Grundzüge der theoretischen Logik*, Berlin: Springer.

David Hilbert and Paul Bernays, 1934, 1939. *Grundlagen der Mathematik*, 2 vols., Berlin: Springer.

Klaus Mainzer, 1988. *Symmetrien der Natur. Ein Handbuch zur Natur- und Wissenschaftsphilosophie*, Berlin: De Gruyter.

Jagdish Mehra, 1974. *Einstein, Hilbert, and the Theory of Gravitation*, Dordrecht: Reidel.

Herbert Mehrtens, 1990. *Moderne—Sprache—Mathematik. Eine Geschichte des Streits um die Grundlagen der Disziplin und des Subjekts formaler Systeme*, Frankfurt am Main: Suhrkamp.

Gregory H. Moore, 1991. "Sixty Years after Gödel," *Mathematical Intelligencer*, 13(3): 6-11.

Abraham Pais, 1982. *'Subtle is the Lord...': The Science and the Life of Albert Einstein*, Oxford: Oxford University Press.

Volker Peckhaus, 1990. *Hilbertprogramm und Kritische Philosophie*, Studien zur Wissenschafts-, Sozial- und Bildungsgeschichte der Mathematik, Bd. 7, Göttingen: Vandenhoeck & Ruprecht.

Michael-Markus Toepell, 1986. *Über die Entstehung von David Hilberts "Grundlagen der Geometrie"*, Studien zur Wissenschafts-, Sozial- und Bildungsgeschichte der Mathematik, Bd. 2, Göttingen: Vandenhoeck & Ruprecht.

Thomas Tymoczko, ed. 1986. *New Directions in the Philosophy of Mathematics*, Boston: Birkhäuser.

Hermann Weyl, 1918. *Raum, Zeit, Materie*, 1st ed., Berlin: Springer.

——— 1944. "David Hilbert and his Mathematical Work," *Bulletin of the American Mathematical Society*, 50: 612-654.

Alfred North Whitehead, 1933. *Science and the Modern World*, Cambridge: Cambridge University Press.

NATUR UND MATHEMATISCHES ERKENNEN

Vorlesungen von David Hilbert,

Gehalten 1919-1920 in Göttingen

ERSTER TEIL:

DIE ÜBLICHE AUFFASSUNG VON DER MATHEMATIK UND IHRE WIDERLEGUNG

1. KAPITEL: DIE ROLLE VON ANSCHAUUNG UND ERFAHRUNG

Wenn wir den ganzen Bereich derjenigen Wissenschaften überblicken, die auf der Betrachtung der leblosen Natur beruhen, dann werden wir vor allem zu denken haben an alle die Tatsachen, die Ergebnisse und Begriffsbildungen, welche den Gebieten der Mathematik, Physik, Chemie, Astronomie, der beschreibenden Naturwissenschaften und der Technik angehören. Dieser Bereich des Wissens ist ein ganz enorm großer. Er grenzt sich gut ab gegenüber dem biologischen Wissen und demjenigen, welches den Menschen als solchen betrifft. Als die charakteristischen Merkmale, durch welche dieses Wissen ausgezeichnet ist, kann man folgende anführen: Erstens ist es älter als die anderen Wissenschaften und daher auch entwickelter, zweitens sind die darin enthaltenen Ergebnisse im großen und ganzen sicherer und damit hängt drittens zusammen, daß sie systematischer geordnet sind.

Es scheint mir daher eine Beschäftigung mit diesem Bereich der Wissenschaft, welchen man ja als die exakte Wissenschaft zu bezeichnen pflegt, besonders wichtig zu sein. Wer innerhalb dieses Gebietes ein Fachstudium betreibt, kommt meist nur dazu, einen kleinen Teil davon kennenzulernen, und da ist es ein Bedürfnis, daß bei dieser Detailforschung der Überblick über das Ganze nicht verloren gehe. Auch ist es ja überhaupt für einen jeden wünschenswert, einen Einblick in die Werkstätte des exakten Forschers zu bekommen und die allgemeinen Gesichtspunkte zum Verständnis der Methode und der Bedeutung exakter Wissenschaft zu gewinnen.

Es ist natürlich nicht möglich, im Rahmen einer Vorlesung den gesamten hier in Betracht kommenden Fragenkreis erschöpfend zu behandeln: Wir müssen vielmehr eine beschränkende Auswahl treffen. Ein regulierendes Prinzip für diese Auswahl finden wir in dem Gedanken, daß es eine der höchsten Aufgaben für die Philosophie ist, die Fragen zu untersuchen, wie Wissen zustandekommt, was Wissen ist und bezweckt. Im Hinblick auf diese Fragen will ich den Stoff behandeln. Meine Vorlesung soll also eine Art Vorbereitung für eine Erkenntnistheorie bilden. Dazu bedarf es, daß wir Kenntnis nehmen von den verschiedenen Sorten der Erkenntnis, sowohl von den scheinbar trivialsten Wahrheiten wie von den jüngsten Ergebnissen wissenschaftlicher Forschung. Entscheidend für die Berücksichtigung der einzelnen Tatsachen soll für uns sein, inwieweit sie beitragen zur Gewinnung einer befriedigenden Einsicht in das Wesen der Erkenntnis selbst.

Die Art der Vorführung wird anders sein als in anderen Vorlesungen; insbesondere kann die Anordnung nicht nach der Schwierigkeit und nicht nach Fächern stattfinden. (Das Wissen als solches kennt keine Fächer.) Unsere Erörterung der Denkmethoden der exakten Wissenschaften werden wir naturgemäß mit der Betrachtung der Mathematik beginnen. Wir werden eine Orientierung über unser Thema anhand der Mathematik versuchen. Dieser Vorsatz bedeutet auch keine Neuerung; die Philosophen sind bereits in dieser Weise verfahren. Ich erinnere an Plato und Kant. Seit den Zeiten Kants hat sich nun ein ungeheures Material an Untersuchungen über die Prinzipien der Mathematik angehäuft. Wir können daher unsere Überlegungen mit weit größerem Erfolge durchführen.

Ich beginne damit, die naheliegende und verbreitete populäre Meinung über Mathematik und mathematisches Denken wiederzugeben. Diese läßt sich so aussprechen: "Die mathematischen Wahrheiten sind absolut sicher; denn sie werden auf Grund von Definitionen durch untrügliche Schlüsse bewiesen. Sie müssen daher auch überall in der Wirklichkeit stimmen; und das tun sie auch." Hiernach wäre die Mathematik streng genommen nur eine ungeheure Tautologie. Alle Sätze würden—wie es Poincaré formuliert hat—nichts anderes lehren als auf Umwegen zu sagen, daß $a = a$ ist. In philosophischer Sprache können wir dafür sagen, daß die Urteile der Mathematik analytische Urteile sind.

Um zu erläutern, wie dies zu verstehen ist, nehme ich als Beispiel den Satz

$$2 + 2 = 4.$$

Um diesen zu beweisen, definiere ich, was 2, was 3, und was 4 ist, durch die Gleichungen:

$$1 + 1 = 2, \quad 2 + 1 = 3, \quad 3 + 1 = 4.$$

Ferner erkläre ich, was "+2" bedeutet durch die Gleichung

$$x + 2 = (x + 1) + 1.$$

Nun kann ich so schließen:

$$2 + 2 = (2 + 1) + 1 = 3 + 1 = 4.$$

Die mathematische Sätze sind hiernach eigentlich Selbstverständlichkeiten: Andererseits sind sie doch wegen der ihnen innewohnenden absoluten Wahrheit sehr wertvoll und wegen ihrer Anwendbarkeit auf die Wirklichkeit sehr bedeutsam; und es findet auf diesem Gebiet eine vollständige Harmonie zwischen Denken und Sein statt.

Trotz der Selbstverständlichkeit der mathematischen Urteile gibt es übrigens, wie Sie wissen, in der Mathematik sehr schwierige Probleme. Ich nenne als Beispiele folgende beiden ungelösten Fragen:

Ist die Gleichung $p - q = 2$ auf unendlich viele Weisen durch Primzahlen p, q lösbar?

Ist $2^{\sqrt{2}}$ eine irrationale Zahl?

Nun fällt uns aber bei näherem Zusehen eine ganze Reihe von Gedanken ein, die uns an der dargelegten Auffassung von der Mathematik stutzig machen. Diese Überlegungen verdichten sich zu strengen Einwendungen, und so werden wir dazu geführt, diese Auffassung von der Mathematik wesentlich zu modifizieren. Und zwar sind es drei Gesichtspunkte, von denen aus die Modifikation zu erfolgen hat.

Die erste Bemerkung, die wir zu machen haben, betrifft folgendes. Wäre die dargelegte Ansicht zutreffend, so müßte die Mathematik nichts anderes als eine Anhäufung von übereinander getürmten logischen Schlüssen sein. Es müßte ein wahlloses Aneinanderreihen von Folgerungen stattfinden, bei welchem das logische Schließen allein die treibende Kraft wäre. Von einer solchen Willkür ist aber tatsächlich keine Rede; vielmehr zeigt sich, daß die Begriffsbildungen in der Mathematik beständig durch Anschauung und Erfahrung geleitet werden, so daß im großen und ganzen die Mathematik ein willkürfreies, geschlossenes Gebilde darstellt. Die Beispiele hierfür sind sehr mannigfaltig. Ich will hier des Näheren das Beispiel der *Begriffsbildung des Kontinuums* behandeln.

Der Begriff des Kontinuums geht aus von der Vorstellung des Fließens, der allmählichen (stetigen) Änderung. Das einfachste anschauliche Bild des Kontinuums ist eine Linie. Wie sich nun aus dieser anschaulichen Vorstellung der mathematische Begriff des Kontinuums entwickelt, wollen wir jetzt genauer betrachten, indem wir die einzelnen Etappen dieser Entwicklung feststellen.

Wir betrachten eine stetige Reihe von Wahrnehmungen, die etwa in der Weise erzeugt werden, daß man mit einer Nadelspitze auf der Haut entlang fährt. Ist die Verschiebung der Nadelspitze genügend klein, so merkt man keinen Unterschied zwischen Anfangs- und Endstellung dieser Verschiebung. Die allmähliche Änderung ist für die Wahrnehmung also dadurch gekennzeichnet, daß in der zeitlichen Aufeinanderfolge zwischen zwei merklich verschiedenen Stellungen A und C der Nadelspitze sich Stellungen B befinden,

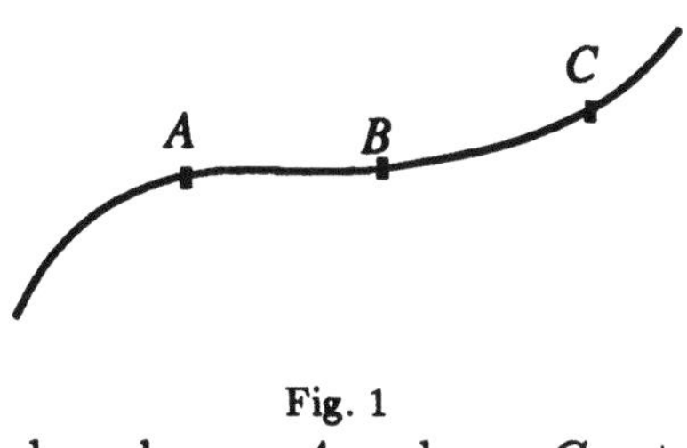

Fig. 1

die sich dem Gefühl nach weder von A noch von C unterscheiden. Soweit wir uns also nur an die Empfindung halten, würde sich der Sachverhalt schematisch durch die Formeln:

$$A = B, \quad B = C, \quad A \neq C$$

ausdrücken lassen.

Diese Formeln enthalten aber nach den Gesetzen der Mathematik einen Widerspruch. Um ihn zu lösen, können wir den Gesichtssinn zu Hilfe nehmen und ihn dabei, falls er allein nicht ausreicht, noch durch ein Vergrößerungsglas unterstützen. Dadurch können wir erreichen, daß die Stellung B wenigstens von einer der beiden Stellungen A und C unterschieden werden kann, so daß der Befund der Wahrnehmungen nicht mehr den Gleichungen $A = B$, $B = C$ entspricht. Der Widerspruch wird damit in der Tat aufgehoben.

Nehmen wir nun an, es sei A als verschieden von B festgestellt; dann werden, da ja die Bewegung der Nadel eine allmähliche sein soll, zwischen A und B jedenfalls noch Stellen beobachtet. Für die Unterscheidung solcher Stellen besteht aber wiederum eine Grenze der Genauigkeit, und so tritt abermals der Widerspruch auf, sobald wir die Stellungen der Nadel, die wir nicht mehr unterscheiden können, als gleich betrachten. So werden wir dazu gezwungen, immer feinere Hilfsmittel der Beobachtung anzuwenden und in der Reihe der Punkte, durch die wir die stetige Bewegung der Nadel darstellen, immer weitere Punkte einzuschalten, so daß wir schließlich dazu geführt werden, auf die Linie unendlich viele Punkte anzunehmen.

Auch noch durch eine andere, prinzipiell wichtige Betrachtung werden wir zu dem Ergebnis geführt, daß das Kontinuum aus unendlich vielen Punkten besteht. Nehmen wir als Darstellung des Kontinuums eine Strecke. Die Hälfte einer solchen, durch eine Lupe gesehen, die im Verhältnis $2 : 1$ vergrößert, sieht genau so aus wie die ganze Strecke, mit bloßem Auge betrachtet, d. h. es besteht kein grundsätzlicher Unterschied zwischen der ganzen Strecke und ihrer Hälfte. Wenn wir nun dies mathematisch ausdrücken wollen, daß die Hälfte gleich beschaffen ist wie das Ganze, so sind wir bereits genötigt, unendlich viele Punkte auf der Strecke anzunehmen. Denn bei einer endlichen Gesamtheit von Dingen läßt sich niemals ein Teil auf das Ganze umkehrbar eindeutig abbilden. Hierin besteht das klassische durchschlagende Kriterium der Endlichkeit: Der Teil ist nie gleich dem Ganzen. Erst bei unendlichen Gesamtheiten ist die umkehrbar eindeutige Abbildung des Ganzen auf einen Teil möglich. Ein Beispiel einer solchen Abbildung ist die Zuordnung aller ganzen Zahlen zu den geraden Zahlen. Man erhält nämlich eine umkehrbar eindeutige Beziehung, indem man jeder Zahl n ihr Doppeltes, $2n$ zuordnet; in diesem Sinne gibt es also ebensoviel gerade Zahlen, wie es überhaupt ganze Zahlen gibt. Auf der Nichtachtung dieser Tatsache beruht sogar ein berühmter Irrtum von Euklid, indem er als selbstverständlich annimmt, daß der Teil eines Dreiecks kleiner als das ganze Dreieck ist—eine Lücke, die erst in neuerer Zeit (in meinen "Grundlagen") ausgefüllt worden ist.

Durch die Methode der unbegrenzt fortgesetzten Einschaltung von Punkten werden wir zu einem *mathematischen Kontinuum "erster Art"* geführt, das arithmetisch durch das System der rationalen Zahlen m/n repräsentiert wird, in welchem ja zwischen zwei verschiedenen Zahlen immer noch weitere Zahlen liegen. Mit dem Kontinuum erster Art kommen wir nun aber für die Zwecke der begrifflichen Präzisierung der Stetigkeitsvorstellung noch nicht aus. Denken wir etwa an die analytische Geometrie der Ebene. Wollten wir hier unter Zugrundelegung der gewöhnlichen rechtwinkligen Parallelkoordinaten nur Punkte mit rationalen Koordinaten als existierend annehmen, so würden wir mit der Anschauung in Widerstreit geraten. Dies zeigt sich an einfachen Beispielen. Gemäß unserer anschaulichen Vorstellung von einer stetigen Linie muß eine Gerade, die aus dem Innern eines Kreises ins Äußere führt, den Kreis wenigstens einmal schneiden. Danach muß also die durch die Gleichung $y = x$ dargestellte Gerade, welche den Nullpunkt mit dem Punkt $(1,1)$ verbindet, mit dem Einheitskreis (d. h. dem Kreis mit dem Radius 1 um den Nullpunkt), dessen Gleichung $x^2 + y^2 = 1$ ist, einen Schnittpunkt haben. Dies würden wir aber nicht zugeben können, wenn wir nur Punkte mit rationalen Koordinaten als existierend gelten lassen. Nach dieser Auffassung gibt es keinen Schnittpunkt der Geraden $y = x$ mit dem Einheitskreis. Denn für einen solchen müßten die Koordinaten gleichzeitig den Bedingungen

$$x^2 + y^2 = 1 \quad \text{und} \quad y = x$$

genügen, sie müßten also die Werte

$$x = y = \pm\frac{1}{\sqrt{2}}$$

haben; aber diese sind ja nicht rational.

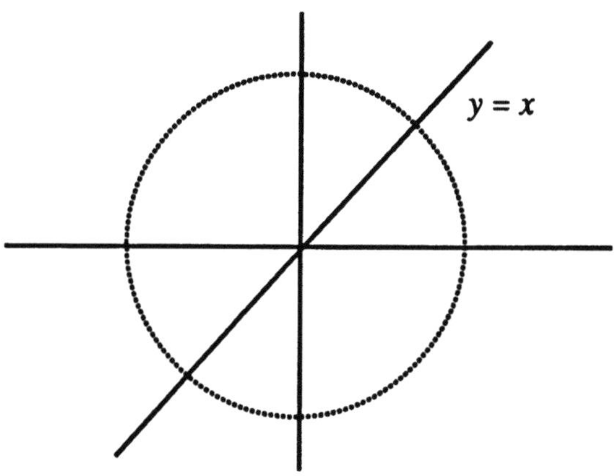

Fig.2

So würden wir also zu dem Schlusse genötigt, daß auf der Geraden $y = x$ kein Punkt im Abstande 1 vom Nullpunkt existiert. Hierdurch kämen wir nicht nur mit der anschaulichen Vorstellung in Widerspruch, sondern auch mit der Elementargeometrie. Denn es ist ja eine der ersten Voraussetzungen der elementaren Geometrie, daß eine gegebenen Strecke (im vorliegenden Falle die Einheitsstrecke) von einem gegebenen Punkt aus auf eine gegebene Gerade abgetragen werden kann.

Wir dürfen also bei dem Kontinuum erster Art nicht stehen bleiben, sondern müssen das System der Zahlen erweitern. Bei dem betrachteten Beispiel würde die Hinzufügung der Zahl $\sqrt{2}$ (und der durch die elementaren Rechenoperationen aus $\sqrt{2}$ zu gewinnenden Zahlen) zur Beseitigung der Schwierigkeit ausreichen. Um aber alle derartigen Fälle mit einem Schlage zu erledigen, bedarf es eines neuen allgemeinen Prinzips der Zwischenschaltung. Ein solches liefert uns das *Dedekindsche Schnittverfahren*, welches in folgendem besteht.

Man betrachtet diejenigen Einteilungen aller rationalen Zahlen in zwei Klassen K und G, bei denen jede Zahl aus K kleiner ist als jede Zahl aus G. Von derartigen Einteilungen gibt es drei typische Fälle, wie sie durch folgende Beispiele dargestellt werden:

1) K bestehe aus den Zahlen, die ≤ 2 sind, G aus denen, welche > 2 sind.

2) K bestehe aus den Zahlen, die < 2 sind, G aus denen, die ≥ 2 sind.

3) K bestehe aus den Zahlen, deren Quadrat < 2 ist, G aus den Zahlen, deren Quadrat > 2 ist.

Bei dem ersten Fall enthält die Klasse K eine größte Zahl, nämlich 2, dagegen enthält G keine kleinste Zahl. Beim zweiten Fall enthält K keine größte Zahl, aber G enthält eine kleinste, nämlich wiederum 2. In diesen beiden Fällen erzeugt gewissermaßen die Zahl 2 die Einteilung. Bei dem dritten Beispiel gibt es weder eine größte Zahl in K noch eine kleinste in G. In diesem Falle nun, wo keine rationale Zahl die Einteilung erzeugt, fassen wir die Einteilung selbst als Definition einer "irrationalen" Zahl auf, welche größer ist als alle Zahlen von K und kleiner als alle Zahlen von G.

So verfahren wir nun bei jeder möglichen Einteilung, wo in K keine größte und in G keine kleinste Zahl existiert. Durch dieses neue Prinzip der Zwischenschaltung von Zahlen gelangen wir von dem System der rationalen Zahlen zu dem System der "reellen" Zahlen, welches ein *Kontinuum zweiter Art* darstellt. Auf diesen erweiterten Zahlenbereich lassen sich die Rechenoperationen der Addition, Subtraktion, Multiplikation und Division derart ausdehnen, daß die gewöhnlichen Rechenregeln alle erhalten bleiben.

Nun fragt es sich, ob der Prozess der Einschaltung von immer neuen Zahlen hiermit ein Ende hat oder ob er noch weiter fortzusetzen ist. Und zwar ist zunächst die Frage, ob ein Weitergehen in der Einschaltung von Zahlen überhaupt logisch möglich ist, ob es insbesondere möglich ist, Zahlen anzunehmen, die zwischen 0 und den sämtlichen Zahlen $1/2, 1/3, 1/4, \ldots$ (welche sich ja der Null unbegrenzt nähern), gelegen sind.

Dies läßt sich nun in der Tat ganz einwandfrei zeigen. Man kann dazu etwa so verfahren: Man betrachtet als Zahlen die rationalen Funktionen eines Arguments ε mit reellen Koeffizienten:

$$R(\varepsilon) = \frac{a_0 + a_1\varepsilon + \ldots + a_n\varepsilon^n}{b_0 + b_1\varepsilon + \ldots + b_m\varepsilon^m},$$

worunter die bisherigen Zahlen (als Konstanten) mitinbegriffen sind. Dabei setzt man fest, daß eine Zahl $R(\varepsilon)$ positiv heißen soll, wenn die Funktion $R(\varepsilon)$ für alle genügend kleinen, positiven Werte von ε positiv ist. Ferner soll $R(\varepsilon)$ größer als $S(\varepsilon)$ heißen, falls $(R(\varepsilon) - S(\varepsilon))$ eine positive Zahl ist. Danach gilt also bei zwei Zahlen $R(\varepsilon)$, $S(\varepsilon)$ die Ungleichung $R(\varepsilon) > S(\varepsilon)$ oder $R(\varepsilon) < S(\varepsilon)$, je nachdem für hinreichend kleine positive Werte von ε die Differenz $R(\varepsilon) - S(\varepsilon)$ beständig positiv oder beständig negativ ist. Bei diesen Festsetzungen behalten alle Rechengesetze, insbesondere die Regeln für das Rechnen mit Ungleichungen ihre Gültigkeit. Und wir erhalten somit durch eine ganz korrekte mathematische Begriffsbildung ein Kontinuum dritter Art. Dieses besitzt auch die Eigenschaft, daß es darin Zahlen gibt, die zwischen 0 und sämtlichen Zahlen $1/2, 1/3, 1/4, \ldots$ liegen. Nämlich ε selbst ist eine solche Zahl. Denn einerseits ist $\varepsilon > 0$, weil ja $\varepsilon - 0$ für jeden positiven Wert von ε positive ist; andererseits ist z. B. $\varepsilon < 1/2$, weil für alle unterhalb von $1/2$ liegenden positiven Werte von ε die Differenz $\varepsilon - 1/2$ negativ ist, und das Entsprechende gilt für die übrigen Zahlen $1/3, 1/4, \ldots$.

Nun hat allerdings diese Konstruktion eines Kontinuums dritter Art noch einen Mangel. Es tritt hier nämlich wieder der Fall ein, daß eine durch den Nullpunkt gehende gerade Linie den Einheitskreis nicht schneidet. In der Tat hat die Gerade $x = \varepsilon y$ mit dem Einheitskreise keinen Punkt gemeinsam; denn die gemeinsame Lösung der Gleichungen

$$x^2 + y^2 = 1 \quad \text{und} \quad \mathrm{x} = \varepsilon \mathrm{y}$$

ist gegebenen durch

$$x = \frac{\varepsilon}{\sqrt{1+\varepsilon^2}}, \qquad y = \frac{1}{\sqrt{1+\varepsilon^2}},$$

diese Ausdrücke stellen aber keine Zahlen unseres Kontinuums dar.

Doch diesem Übelstand läßt sich abhelfen. Man braucht nur den Bereich der Zahlen etwas weiter zu fassen, indem man sich nicht auf rationale Funktionen von ε beschränkt, sondern alle Ausdrücke der Form

$$a \cdot \varepsilon^r \cdot (1 + a_1\varepsilon^{1/n} + a_2\varepsilon^{2/n} + \ldots)$$

zuläßt, worin r eine rationale Zahl (mit beliebigem Vorzeichen), n eine positive ganze Zahl bedeutet, die Koeffizienten $a, a_1, a_2, \ldots$ reelle Zahlen sind und die Potenzentwicklung $1 + a_1\varepsilon^{1/n} + \ldots$ für hinreichend kleine Werte von ε konvergiert.

Hierdurch wird erreicht, daß in der dem Zahlensystem entsprechenden analytischen Geometrie der Ebene alle im Sinne der anschaulichen Stetigkeit vorhandenen Schnittpunkte von Kurven wirklich auch durch Punkte der Koordinatenebene dargestellt werden. So stellen in den angeführten Beispiel die Ausdrücke

$$\frac{\varepsilon}{\sqrt{1+\varepsilon^2}}\,, \quad \frac{1}{\sqrt{1+\varepsilon^2}}$$

Zahlen unseres erweiterten Systems dar, weil ja

$$\frac{1}{\sqrt{1+\varepsilon^2}}$$

in eine Reihe nach Potenzen von ε entwickelbar ist, die für kleine Werte von ε konvergiert. Und durch diese Zahlen als Koordinaten wird ein Schnittpunkt der Geraden mit dem Einheitskreis bestimmt.

Man sieht nun leicht, daß in dem erweiterten Zahlenbereich alle elementaren Rechenoperationen ausführbar sind, auch läßt sich "größer" und "kleiner" in derselben Weise wie vordem einführen. Die Definition ist am einfachsten so zu formulieren: eine Zahl

$$F(\varepsilon) = a\varepsilon^r(1 + a_1\varepsilon^{1/n} + \ldots)$$

heißt positiv, wenn der Koeffizient a im gewöhnlichen Sinne positiv ist, und eine Zahl $F(\varepsilon)$ heißt größer als $G(\varepsilon)$, wenn die Differenz $F(\varepsilon) - G(\varepsilon)$ positiv ist. Hierbei bleiben wiederum die gewöhnlichen Regeln für das Rechnen alle gültig. So läßt sich also in einer mathematisch ganz einwandfreien Weise der Übergang zu einem Kontinuum dritter Art vollziehen, wobei man es einrichten kann, daß die bei dem Kontinuum erster Art vorgefundenen Mängel hier nicht wieder auftreten.

Sehen wir nun zu, was die Annahme eines Kontinuums dritter Art für Folgen haben würde. Es gibt einen mathematischen Grundsatz, welchen man als *Archimedische Forderung* bezeichnet. Dieser besagt folgendes: Es sei auf einer Geraden eine Strecke $\overline{AB}$ und ein Punkt P außerhalb der Strecke gegeben. Wenn ich dann die Strecke $\overline{AB}$ hinreichend oft hintereinander in der Richtung nach P hin abtrage, so komme ich jedenfalls einmal über P hinaus. Die Strecke $\overline{AB}$ mag noch so klein sein, und der Punkt P mag noch so weit entfernt liegen (etwa so weit wie der Sirius von der Erde), so wird doch nach einer endlichen Zahl von Abtragungen der Punkt P überschritten.

Fig. 3

Dieses Archimedische Postulat ist bei dem Kontinuum dritter Art nicht erfüllt. Denn wählen wir z. B. als Punkte A, B, P die Punkte $0, 1, 1/\varepsilon$ auf der Zahlenachse, so kann der Punkt P durch noch so vielmaliges Abtragen der Einheitsstrecke nicht erreicht werden, weil jede ganze Zahl n kleiner ist als $1/\varepsilon$. (Es gelten ja für alle positiven Werte von ε, die kleiner als $1/n$ sind, die Ungleichungen $n\varepsilon < 1, \quad n < 1/\varepsilon$.) Ebenso ergibt sich auch, daß der Punkt 1 durch noch so vielmaliges Abtragen der Strecke ε vom Nullpunkt aus nicht erreicht werden kann.

Nun werden wir aber diese Forderung ohne zwingenden Anlaß nicht preisgeben. Und ein solcher Anlaß liegt in unserer Erfahrung auch gar nicht vor, vielmehr besteht die bemerkenswerte Tatsache, daß alle Abstände in der Natur miteinander zahlenmäßig vergleichbar sind. Die größten astronomischen Entfernungen sowie die kleinsten Abstände, wie wir sie bei den Atomen und Elektronen finden, lassen sich alle durch Vielfache und Bruchteile eines Zentimeters ausdrücken. Daher ist das Kontinuum dritter Art, welches als Zahlensystem in der Mathematik natürlich seine Bedeutung hat, für den Zweck der mathematischen Auffassung des Stetigen abzulehnen, und wir bleiben also bei dem Kontinuum zweiter Art stehen.

So sehen wir an dem Beispiel des Kontinuums, wie die mathematische Begriffsbildung durch die Anschauung angeregt und von der Erfahrung geleitet wird. Es ließen sich hierfür noch zahlreiche Beispiele geben. Insbesondere ist zu erwähnen, daß der Anstoß zur Ausbildung der Infinitesimalrechnung von der Geometrie und Mechanik ausgegangen ist, und daß deren weitere Entwicklung durch die Probleme der Naturwissenschaft und der Technik wesentlich mitbestimmt worden ist und noch wird.

Das vorige Mal habe ich auseinandergesetzt, daß entgegen der Ansicht, wonach die Mathematik ganz wahllos ihre Schlüsse anhäuft, die mathematische Wissenschaft im großen und ganzen ein willkürfreies geschlossenes System darstellt. Insbesondere zeigte ich am Beispiel des Kontinuums, wie die Begriffsbildung der Mathematik, indem sie sich an die Anschauung und die Erfahrungstatsachen hält, auf eine ganz bestimmte Weise gelenkt wird, so daß von Willkür gar keine Rede ist. Ich wies auch darauf hin, daß die Anforderungen der Physik und der Technik auf den Entwicklungsgang der Mathematik richtunggebend einwirken.

Zu dieser Lenkung von außen her, welche die Mathematik erfährt, kommt nun aber noch eine innere Tendenz der Mathematik zur Systematik, welche teils in einem bewußten Streben nach logischer Geschlossenheit besteht und teils darauf beruht, daß die Verfolgung der naheliegenden mathematischen

Fragestellungen von selbst zu einer systematischen Ausbildung der Wissenschaft nötigt. Auch bei dem Falle des Kontinuums geschieht es nicht nur im Hinblick auf die Darstellung der räumlichen und physikalischen Verhältnisse, daß wir uns für das Kontinuum zweiter Art entscheiden, sondern es spricht auch ein innerer Grund dafür, welcher der Betrachtungsweise der mathematischen Logik angehört.

Es handelt sich dabei um den Gesichtspunkt der ***Vollständigkeit*** des Zahlensystems. Ein System von Dingen, zwischen denen gewisse Relationen bestehen, heißt mit Bezug auf bestimmte Eigenschaften dieser Relationen ein vollständiges, wenn es bei Aufrechterhaltung dieser Eigenschaften keiner Erweiterung mehr fähig ist.

Bei dem System der Zahlen des Kontinuums, welche durch die elementaren Rechenoperationen und durch die Beziehung des "größer" und "kleiner" verknüpft sind, halten wir als Eigenschaften jedenfalls die gewöhnlichen Gesetze des Rechnens fest. Dazu kommt noch die Eigenschaft, welche durch die Archimedische Forderung ausgedrückt wird. Alle diese Eigenschaften besitzt das Kontinuum erster Art, aber es ist in Hinsicht auf diese kein vollständiges System; denn es läßt sich ja zu dem Kontinuum zweiter Art erweitern, bei welchem auch noch die Rechengesetze und die Archimedische Forderung erfüllt sind.

Das Kontinuum zweiter Art aber genügt der Vollständigkeitsbedingung. Denn ließe es sich unter Aufrechterhaltung der Rechenregeln und der Archimedischen Forderung zu einem umfassenderen System erweitern, so gäbe es in diesem eine Zahl a, welche nicht zu den reellen Zahlen gehörte. Diese müßte mit allen reellen Zahlen, und insbesondere auch mit allen rationalen Zahlen der Größe nach vergleichbar sein, und zwar könnte sie auf Grund der Archimedischen Forderung nicht kleiner sein als sämtliche rationalen Zahlen und ebensowenig größer als alle rationalen Zahlen. Daher würde die Einteilung der rationalen Zahlen in solche, die kleiner als a (Klasse K) und solche, die größer als a (Klasse G) sind, einen Dedekindschen Schnitt darstellen. Diesem Schnitte müßte eine reelle Zahl b entsprechen, welche (gleichgültig, ob sie rational oder irrational ist) kleiner oder gleich jeder Zahl von G und größer oder gleich jeder Zahl von K ist. Da nun aber a keine reelle Zahl sein soll, so müßte b von a verschieden sein, und zufolge der Archimedischen Forderung müßten daher zwischen b und a rationale Zahlen liegen. Es wäre demnach entweder b kleiner als gewisse Zahlen aus K oder gar als gewisse Zahlen aus G, und wir kämen also auf einen Widerspruch. Somit erkennen wir, daß bei Zugrundelegung der gewöhnlichen Rechenregeln sowie der Archimedischen Forderung das Kontinuum zweiter Art durch die Bedingung der Vollständigkeit—welche ich als "Vollständigkeitsaxiom" aufgestellt habe—, eindeutig festgelegt ist. Eine Erweiterung des Kontinuums zweiter Art bei Festhaltung der üblichen Rechengesetze ist nur dadurch möglich, daß man die Archimedische Forderung fallen läßt, und es ergibt sich hiermit ganz allgemein, daß ein Kontinuum dritter Art der Archimedischen Forderung nicht genügen kann.

Wenn man nun so verfährt, daß man die Archimedische Forderung preisgibt und nur die Rechengesetze als Eigenschaften des Zahlensystems festlegt, dann ist tatsächlich, wie wir gesehen haben, das Kontinuum zweiter Art nicht mehr vollständig sondern vielmehr einer Erweiterung fähig. Man könnte nun denken, daß sich durch Hinzufügung hinreichend vieler Zahlen wiederum ein vollständiges System herstellen ließe. Das ist jedoch nicht der Fall. Man kann nämlich die Methode, mit deren Hilfe wir das Kontinuum zweiter Art erweitert haben, auf jedes System S anwenden, in welchem die Gesetze für die elementaren Rechenoperationen und für die Ungleichheitsbeziehungen gelten. Man betrachtet als Zahlen im weiteren Sinne die rationalen Funktionen eines Arguments w, deren Koeffizienten Zahlen von S sind; eine solche Zahl $R(w)$ nennt man positiv, wenn bei der Entwicklung von $R(w)$ nach steigenden Potenzen von w der erste Koeffizient eine positive Zahl von S ist, und man nennt $R(w)$ größer als $S(w)$, wenn $R(w) - S(w)$ positiv ist. So erhält man ein System, in welchem S als Teilsystem enthalten ist und worin alle Rechenregeln gültig bleiben. Es ist also beim Aufgeben der Archimedischen Forderung nicht mehr möglich, dem Vollständigkeitsaxiom zu genügen. Wir können ein vollständiges Zahlensystem nur in dem Sinne erreichen, daß wir die Erfüllung der Archimedischen Forderung mit zu den charakteristischen Eigenschaften des Zahlensystems rechnen. Tun wir dies aber, so werden wir durch die Bedingung der Vollständigkeit gerade zu dem Kontinuum zweiter Art geführt; in diesem findet somit die Zahlenbildung ihren natürlichen Abschluß. Wir sehen also, daß das Kontinuum zweiter Art sowohl gegenüber den engeren wie auch gegenüber den weiteren Zahlensystemen einen inneren Vorzug besitzt, und daß daher auch vom logisch-systematischen Standpunkt ein Grund dafür besteht, bei der Begriffsbildung des Kontinuums gerade bis zu dem Kontinuum zweiter Art zu gehen und bei diesem stehenzubleiben.

Ebenso wie an diesem speziellen Beispiel zeigt sich überhaupt bei einer näheren Betrachtung der mathematischen Theorien, daß die Entwicklung der Mathematik, abgesehen von dem Einfluß, der von den Bedürfnissen der Geometrie und der Physik ausgeht, schon bei der Art, wie sie durch das reine Denken fortschreitet, im wesentlichen eine zwangsläufige ist.

Den Ausgang bilden gewisse naheliegende Probleme, insbesondere aus der Gleichungstheorie und der Zahlentheorie, auf welche man bei Beschäftigung mit den Zahlen ohne weiteres geführt wird (z. B. die Frage, wann eine ganze Zahl sich als Summe von zwei Quadratzahlen darstellen läßt). Zur Lösung solcher Probleme sind nun ganz betimmte Begriffsbildungen und Beweismethoden nötig. Freilich kann ein einzelner Satz oft auf verschiedene Weisen bewiesen werden; aber diese verschiedenen Möglichkeiten des Beweises wird man dann auch zur vollen Klärung der Zusammenhänge alle aufsuchen. Im ganzen ist der Gang der Untersuchungen beinahe eindeutig vorgezeichnet, so daß man keinen wesentlich anderen Weg einschlagen kann als den, welchen die Wissenschaft tatsächlich genommen hat.

Dies gilt nicht nur für die Ausbildung der Schlußweise innerhalb einer Theorie, sondern auch für die Erschließung neuer Gebiete des mathematischen Wissens. Die Entwicklung der Methoden der Infinitesimalrechnung und der Funktionentheorie ist vom Standpunkt des reinen mathematischen Denkens keineswegs zufällig, vielmehr erweisen sich die begrifflichen Konstruktionen und die Schlußweisen der Analysis in dem Fortgang der zahlentheoretischen und algebraischen Untersuchungen als unentbehrliche Hilfsmittel zur Behandlung der Probleme. So ist für die Probleme, welche sich an die reinen Gleichungen (d. h. Gleichungen der Form $x^n = 1$) knüpfen, die Exponentialfunktion von ausschlaggebender Bedeutung; und bei den höheren zahlentheoretischen Fragen wird es notwendig, die Theorie der elliptischen Funktionen und der elliptischen Modulfunktionen heranzuziehen. Auch die Fragen, welche die Verteilung der Primzahlen betreffen, erforden zu ihrer Lösung die transzendenten Hilfsmittel der Funktionentheorie.

Es bilden also die verschiedenen vorliegenden mathematischen Disziplinen notwendige Glieder im Aufbau einer systematischen Gedankenentwicklung, welche von einfachen, naturgemäß sich bietenden Fragen anhebend, auf einem durch den Zwang innerer Gründe im wesentlichen vorgezeichneten Wege fortschreitet. Von Willkür ist hier keine Rede. Die Mathematik ist nicht wie ein Spiel, bei dem die Aufgaben durch willkürlich erdachte Regeln bestimmt werden, sondern ein begriffliches System von innerer Notwendigkeit, das nur so und nicht anders sein kann.

2. KAPITEL: DIE ROLLE DER VORAUSSETZUNGEN

Ich komme nun zu dem zweiten Gesichtspunkt, unter welchem die landläufige Auffassung von der Mathematik einer Korrektur bedarf. Man sagt da, in der Mathematik werden alle Sätze bewiesen. Danach würde also die Mathematik eine gänzlich voraussetzungslose Wissenschaft sein, in der es keinerlei prinzipielle Schwierigkeiten gibt. Die tatsächlichen Schwierigkeiten bei den Beweisen, von denen ja die Rede war, würden nur darauf beruhen, daß wir oft die für die Schlüsse nötigen Gedankenkombinationen nicht finden.

Wenn man sich das näher überlegt, dann sieht man gleich, daß diese Meinung jedenfalls nicht so ohne weiteres zutreffen kann. Denn aus nichts kann auch nichts bewiesen werden. Es muß also bei dem logischen Schließen ein Rest von Annahmen übrig bleiben; es müssen Voraussetzungen da sein, auf die wir uns bei den Beweisen stützen. Und wenn wir uns die mathematischen Disziplinen ansehen, so finden wir da auch tatsächlich, daß gewisse letzte Voraussetzungen die Grundlage für die Beweise bilden.

Hierdurch scheint die Mathematik den Charakter des absolut Sicheren zu verlieren. Denn mit der Annahme von Voraussetzungen kommen wir auf das Gebiet des Problematischen. Beruhen doch die Meinungsverschiedenheiten der Menschen zumeist darauf, daß von verschiedenen Voraussetzungen ausgegangen wird. Nun wird man aber sagen: Die Mathematik hat zwar bestimmte Grundannahmen; aber diese sind so selbstverständlich, so ohne weiteres einleuchtend, daß sie dem Streit der Meinungen entzogen sind. Man wird also den Grund für den Erfolg und die Sicherheit der mathematischen Wissenschaft darin sehen, daß sie nur ganz einleuchtende Annahmen zugrundelegt, über die gar nicht gestritten werden kann.

Es ist dies eine Vorstellung, die sich sehr eingebürgert hat. Man verweist dabei insbesondere auf die Euklidische Geometrie, die ein klassisches Beispiel logischer Systematik bildet. In dieser werden die Grundannahmen ausdrücklich als "Axiome" an die Spitze gestellt, und zwar sind diese Axiome so gewählt, daß sie den Charakter anschaulicher Klarheit, des unmittelbar Einleuchtenden besitzen.

Dies Verfahren ist von jeher als ein Muster strenger Wissenschaftlichkeit angesehen worden und hat daher vielen anderen Wissenschaften als Vorbild gedient. Insbesondere in der theoretischen Physik war man bestrebt, sich dem Ideal der Euklidischen Geometrie zu nähern. Man suchte den Theorien dadurch den Charakter Sicherheit zu verleihen, daß man einige wenige, unmittelbar einleuchtende Sätze als Prinzipien an die Spitze stellte. Beispiele für solche einleuchtenden Prinzipien sind Newtons Grundsätze der Mechanik: der Satz von der Erhaltung der Masse, das Trägheitsgesetz und der Satz über die Gleichheit von Wirkung und Gegenwirkung;* sodann in der Wärmelehre das Prinzip von Clausius, daß der Wärmestrom von selbst nur von einem wärmeren zu einem kälteren Körper überfließt (aber nicht umgekehrt).

Der Gesichtspunkt des Einleuchtenden bestimmte auch Heinrich Hertz bei seiner Grundlegung der Mechanik. Den Begriff der Kraft, der ihm als zu unanschaulich erschien, lehnte er ab, und es ist ihm gelungen, die Prinzipien der Mechanik auf eine solche Form zu bringen, daß darin nur von Massen, ihren Bewegungen (Geschwindigkeiten, Beschleunigungen) und von festen Verbindungen zwischen den Massen die Rede ist. Hier überall betrachtet man es als ein Erfordernis der exakten Wissenschaft, daß ihre Grundsätze als wahr einleuchten, damit sich von ihnen die Sicherheit vermittels der logischen Schlüsse auf die Lehrsätze der Theorie überträgt.

Dieser Standpunkt stellt eine erste Modifikation des Gedankens von der Voraussetzungslosigkeit der Mathematik dar. Nun müssen wir aber auch gegen diesen Bedenken erheben. Der Gesichtspunkt des Einleuchtenden hat nämlich das Mißliche, daß keine volle Einhelligkeit darüber besteht, was als selbstverständlich zu gelten hat. Oftmals ist ein Fortschritt unserer Erkenntnis gerade dadurch zustandegekommen, daß man sich davon überzeugt hat, daß etwas, was man für selbstverständlich hielt, in Wahrheit nicht selbstverständlich ist. In den Voraussetzungen, die als unmittelbar einleuchtend erscheinen, sind mitunter Vorurteile enthalten, und die Überwindung eines solchen Vorurteils kann einen wichtigen Fortschritt für die Wissenschaft bedeuten. In der Betonung dieses Gedankens geht ein neuerer Philosoph so weit, daß er die Maxime aufstellt: "Beargwöhne jeden Satz als falsch, der dir evident, denknotwendig und unbezweifelbar erscheint."

Wie wenig man sich auf den Charakter des Einleuchtenden bei einem Grundsatz verlassen kann, das hat insbesondere die neuere Entwicklung der Physik gelehrt. Hier haben Voraussetzungen, die man für vollkommen selbstverständlich hielt, preisgegeben werden müssen. So müßte z. B. der Satz vom Parallelogramm der Bewegungen fallen gelassen werden, und auch das Trägheitsgesetz hat man in seiner ursprünglichen Bedeutung nicht aufrechterhalten können, indem sich (bei der Beobachtung der Elektronenbewegung) die Trägheit als abhängig von der Geschwindigkeit erwies.

Glücklicherweise verhält es sich nun so, daß für den Erfolg der mathematischen Methode die Sicherheit der Grundannahme gar kein Erfordernis bildet. In der mathematischen Physik ist es ja ein ganz bekanntes und geläufiges Verfahren, daß man von Annahmen ausgeht, die noch gar nicht feststehen, von denen vielmehr erst der Erfolg lehren soll, ob sie zutreffen.

---

* Die Gültigkeit dieser Prinzipien suchte Kant (in den "Metaphysischen Anfangsgründen der Naturwissenschaft") aus philosophischen Gründen zu erweisen.

Hierbei sind sich die Physiker dessen ganz bewußt, daß sie die Voraussetzungen ihrer Theorien nur versuchsweise zugrundelegen, und mitunter wird es auch notwendig, diese Voraussetzungen zu ändern. Die Wahl der Axiome geschieht keineswegs immer nach dem Gesichtspunkt des Einleuchtenden. Speziell die moderne theoretische Physik ist für die Auffassung, daß man als Grundlage das Einleuchtende ansieht, geradezu ein schlagendes Gegenbeispiel. Die neuen physikalischen Theorien verlangen eine Auffassung von Raum und Zeit, welche von der uns anschaulich geläufigen grundsätzlich verschieden ist. Hier geht die Abweichung von dem Einleuchtenden so weit, daß man eine große geistige Anstrengung anwenden muß, um sich von dem anschaulich Vertrauten unabhängig zu machen. Auf diese neue Art der Grundlegung würden wir niemals gekommen sein, wenn uns nicht die Experimente darauf gestoßen hätten.

Obwohl nun die Grundannahmen der neueren theoretischen Physik durchaus nicht einleuchtend, geschweige denn selbstverständlich sind, so bewährt die mathematische Methode sich gerade hier aufs glänzendste. Worauf beruht denn nun eigentlich der Erfolg des mathematischen Denkens? Die Antwort lautet: Das Wesentliche der mathematischen Methode besteht in der konsequenten Ausgestaltung des Verfahrens, das dem formalen Denken eigentümlich ist, also des logischen Denkens überhaupt. Dem Denken steht ein unermeßlicher Vorrat an formalen Beziehungen zur Verfügung, und es kommt darauf an, solche Systeme von formalen Beziehungen zu finden, die sich den in der Wirklichkeit vorgefundenen Beziehungen anpassen lassen.

Die Wirklichkeit wird also daraufhin untersucht, wie weit sie sich durch begriffliche Relationssysteme darstellen läßt. Von Physikern ist es ja oft hervorgehoben worden: Die Begriffssysteme sollen Abbilder der Beziehungen sein, die sich in der Wirklichkeit vorfinden. Dementsprechend kann auch in der mathematischen Naturwissenschaft von der Wahrnehmung nur das verwendet werden, was sich durch begriffliche Beziehungen deuten läßt, worüber also eine von der Wahl der Sinne unabhängige Verständigung möglich ist. Es werden letzten Endes nur solche Wahrnehmungen benutzt, die in der Feststellung einer Koinzidenz (z. B. Zusammenfallen eines Zeigers mit einem Teilstrich einer Skala) bestehen, da hierüber eine objektive Verständigung am ehesten zu erreichen ist.

Der Mathematik kommt hierbei eine zweifache Aufgabe zu: Einerseits gilt es, die Systeme von Relationen zu entwickeln und auf ihre logischen Konsequenzen zu untersuchen, wie dies ja in den rein mathematischen Disziplinen geschieht. Dies ist die *progressive Aufgabe* der Mathematik. Andererseits kommt es darauf an, den an Hand der Erfahrung gebildeten Theorien ein festeres Gefüge und eine möglichst einfache Grundlage zu geben. Hierzu ist es nötig, die Voraussetzungen deutlich herauszuarbeiten, und überall genau zu unterscheiden, was Annahme und was logische Folgerung ist. Dadurch gewinnt man insbesondere auch Klarheit über alle unbewußt gemachten Voraussetzungen, und man erkennt die Tragweite der verschiedenen Annahmen, so daß man übersehen kann, was für Modifikationen sich ergeben, falls eine oder die andere von diesen Annahmen aufgehoben werden muß. Dies ist die *regressive Aufgabe* der Mathematik.

Diese beiden Aufgaben des mathematischen Denkens sind von sehr allgemeiner Bedeutung; sie beziehen sich nicht nur auf den Kreis der Naturwissenschaften, sondern sie gelten auch für andere Wissensgebiete, z. B. für die Nationalökonomie (Theorie des Geldes). Auch in der Philosophie wird verschiedentlich versucht, das mathematische Denken zur Geltung zu bringen. So ahmt Spinoza in seinem Hauptwerk, der "Ethik", die progressive Methode nach, während neuerdings Nelson in seiner Philosophie von der regressiven Methode der Mathematik Gebrauch macht.

Diese regressive Methode findet ihren vollkommensten Ausdruck in dem, was man heute die "axiomatische Methode" nennt. Diese bildet eine allgemeine Methode des wissenschaftlichen Forschens überhaupt; ihre glänzendsten Triumphe feiert sie aber in der Mathematik. Zur näheren Charakterisierung der axiomatischen Methode möchte ich folgendes sagen: Wenn wir irgendein Gebiet wissenschaftlich untersuchen wollen, so werden wir zunächst die vorhandenen Tatsachen dieser Disziplin sammeln. Wir werden dann bemerken, daß diese Tatsachen einer Ordnung bedürftig und fähig sind. Diese Ordnung erfolgt mit Hilfe eines Fachwerks von Begriffen, und zwar in der Weise, daß dem einzelnen Gegenstande des Wissensgebietes ein Begriff dieses Fachwerks und jeder Tatsache innerhalb des Wissensgebietes eine logische Beziehung zwischen den Begriffen entspricht. Das Fachwerk von Begriffen stellt dann die Theorie des Wissensgebietes dar. Beispiele von Theorien dieser Art sind in der reinen Mathematik die Flächentheorie, die Galois'sche Theorie der Gleichungen, die Theorie der Primzahlen, die Mengenlehre usw.; in der mathematischen Physik die Statik, die geometrische Optik, die Thermodynamik, die Gastheorie.

Nun erkennt man sehr bald, daß einige ausgezeichnete Sätze, deren Anzahl meist sehr gering ist, der Theorie als Grundlage dienen oder dienen können, derart, daß auf dieser Grundlage ohne jede weitere Annahme das ganze Fachwerk mit rein logischen Schlüssen aufgebaut werden kann. Diese grundlegenden Sätze heißen dann die Axiome der betreffenden Theorie. So spielt z. B. in der Galois'schen Gleichungstheorie der Satz von der Existenz der Lösungen einer algebraischen Gleichung, in der Statik der Satz vom Parallelogramm der Kräfte, in der geometrischen Optik der Satz von der geradlinigen Fortpflanzung des Lichtes die Rolle eines Axioms.

Hat man nun ein solches Axiomensystem aufgestellt, so kann man weiter fragen, ob die Axiome nicht ihrerseits (wenigstens zum Teil) beweisbar sind. Diese Frage kann in zweierlei Sinn gemeint sein. Es kann sich darum handeln, zu entscheiden, ob ein Axiom aus den übrigen logisch gefolgert werden kann, ob es also ein überflüssiges Axiom ist. Besonders berühmt ist in dieser Hinsicht die Frage nach der Beweisbarkeit des Parallelenaxioms, welche im verneinenden Sinne entschieden wurde. Aus diesem Problem ist historisch die axiomatische Untersuchungsweise hervorgegangen. Es kann aber auch ein Beweis von Axiomen in dem Sinne gesucht werden, daß man andere Sätze von allgemeinerem Charakter als Axiome wählt. So hat Gauss die Existenz der Wurzeln algebraischer Gleichungen bewiesen, indem er die allgemeinen Grundlagen der Funktionentheorie (die Theorie des mathematischen Kontinuums) benutzte; und der Satz von der geradlinigen Ausbreitung des Lichtes läßt sich (in dem Umfange, in welchem er zutrifft) aus der Wellentheorie des Lichtes ableiten. Der Fortschritt bei solchen Beweisführungen besteht darin, daß die Fundamente der Wissenschaft tiefer gelegt werden; man geht damit in der logischen Zurückführung noch ein Stück weiter. Dieses Tieferlegen der Fundamente wird in dem Maße nötig, als man das Gebäude der Wissenschaft höher ausbaut.* Dies ist das Wesentliche, was zur Berichtigung der Ansicht von der Voraussetzungslosigkeit der Mathematik zu sagen ist.

* Auf die Fragen der Widerspruchslosigkeit von Axiomensystemen will ich hier nicht näher eingehen.

3. KAPITEL: DIE NICHTUNTRÜGLICHKEIT DES MATHEMATISCHEN SCHLIESSENS

In der Berichtigung der landläufigen Auffassung über die Mathematik gehe ich nunmehr zu dem dritten Punkt über. Es handelt sich dabei um das mathematische Schließen selbst. Dieses wird nach der üblichen Meinung für etwas ganz Selbstverständliches angesehen. Man braucht ja, so denkt man, nur die gewöhnlichen Regeln der Logik anzuwenden; dann gelangt man, von den Axiomen ausgehend, zu dem ganzen Aufbau der mathematischen Theorien. Und soweit die Axiome sicher sind, müssen auch die Folgerungen zuverlässig sein. Eine prinzipielle Schwierigkeit kann hiernach also gar nicht auftreten.

Diese Auffassung ist für die elementare Geometrie und Zahlentheorie wohl zutreffend. Doch mit diesen Disziplinen ist ja die mathematische Wissenschaft, und zwar auch die reine Mathematik, nicht erschöpft, es bleibt vor allem das große und wichtige Gebiet der *Infinitesimalrechnung* zu betrachten. Da ist es nun eine historisch bekannte Tatsache, daß in diesem Gebiete der Mathematik die Schlußweisen anfangs höchst bedenklicher Art waren. Durch die Entdeckung der Infinitesimalrechnung wurden der Mathematik ganz ungeahnte Möglichkeiten eröffnet, und ganz überraschende Zusammenhänge traten zutage. Aber es herrschte keine völlige Klarheit über die logische Bedeutung der neuen Methode. Nun war man natürlich zunächst darauf bedacht, in dem neu erschlossenen, fruchtbaren Feld möglichst schnell vorzudringen. So kam es, daß man die Forderung einer strengen logischen Begründung beiseite stellte und sich bei der Wahl der Schlußweisen auf das mathematische Gefühl verließ. Die glänzenden Erfolge, welche die ersten Forscher in der Infinitesimalrechnung aufzuweisen hatten, schienen zunächst auch die Kühnheit ihres Verfahrens zu rechtfertigen. Aber bald stellten sich Widersprüche heraus, durch welche die Sicherheit der Ergebnisse so sehr in Frage gezogen wurde, daß bedeutende Denker, wie z. B. der Philosoph Berkeley, die ganze Methode der Infinitesimalrechnung geradezu als unwissenschaftlich ablehnten. Von solchen Widersprüchen will ich einige vorführen.

Betrachten wir die bei Euler vorkommende unendliche Reihe:

$$1 - 1 + 1 - 1 + 1 - 1 + \dots \quad .$$

Von einer solchen Reihe nahmen die Mathematiker jener Zeit ohne weiteres an, daß sie einen bestimmten Wert a besitze. Um diesen Wert zu ermitteln, bildet man die Differenz $1 - a$ und findet

$$\begin{aligned} 1 - a &= 1 - (1 - 1 + 1 - 1 + \dots) \\ &= 1 - 1 + 1 - 1 + 1 - \dots \\ &= a. \end{aligned}$$

Somit ergibt sich $a = 1/2$. Andererseits aber ist

$$\begin{aligned} a &= (1-1)+(1-1)+(1-1)+\dots \\ &= 0+0+0+\dots, \\ \text{d. h.}\quad a &= 0. \end{aligned}$$

Wir erhalten also zwei widersprechende Ergebnisse.

Ein weiteres Beispiel ist die (nach beiden Seiten ins Unendliche gehende) Reihe

$$\dots + x^4 + x^3 + x^2 + x + 1 + \frac{1}{x} + \frac{1}{x^2} + \frac{1}{x^3} + \dots$$

Diese läßt sich in zwei Teile zerlegen:

$$\frac{1}{x} + \frac{1}{x^2} + \frac{1}{x^3} + \frac{1}{x^4} \dots$$

und

$$1 + x + x^2 + x^3 + \dots \quad .$$

Gemäß der Summenformel für die geometrische Reihe ist der erste Bestandteil gleich

$$\frac{\frac{1}{x}}{1-\frac{1}{x}} = \frac{1}{x-1},$$

der zweite gleich $1/(1-x)$. Da diese beiden Ausdrücke zusammen Null ergeben, so müßte der Wert der vorgelegten Reihe stets gleich Null sein. Das steht aber im Widerspruch damit, daß für positive Werte von x jedes Glied der Reihe größer als Null ist.

Ein Widerspruch von etwas anderer Art ist folgender: In der unendlichen Reihe

$$1 - \frac{1}{2} + \frac{1}{3} - \frac{1}{4} + \frac{1}{5} - \frac{1}{6} + \dots,$$

deren Wert mit s bezeichnet werde, ändere man die Reihenfolge der Glieder in der Weise, daß je auf ein positives Glied zwei negative Glieder folgen. So erhält man

$$\begin{aligned} s &= 1 - \frac{1}{2} - \frac{1}{4} + \frac{1}{3} - \frac{1}{6} - \frac{1}{8} + \frac{1}{5} - \frac{1}{10} - \frac{1}{12} + \dots \\ &= \left(1-\frac{1}{2}\right) - \frac{1}{4} + \left(\frac{1}{3} - \frac{1}{6}\right) - \frac{1}{8} + \left(\frac{1}{5} - \frac{1}{10}\right) - \frac{1}{12} + \dots \\ &= \frac{1}{2} - \frac{1}{4} + \frac{1}{6} - \frac{1}{8} + \frac{1}{10} - \frac{1}{12} + \dots \\ &= \frac{1}{2}\left(1 - \frac{1}{2} + \frac{1}{3} - \frac{1}{4} - + \frac{1}{5} - \frac{1}{6} + \dots\right), \end{aligned}$$

d. h.

$$s = \frac{1}{2s}.$$

Diese Gleichung kann nur bestehen, wenn $s = 0$ ist. Das ist aber unmöglich; denn die Reihe für s läßt sich ja schreiben in der Form

$$\left(1 - \frac{1}{2}\right) + \left(\frac{1}{3} - \frac{1}{4}\right) + \left(\frac{1}{5} - \frac{1}{6}\right) + \dots,$$

wobei in jeder der Klammern ein positiver Ausdruck steht; es muß demnach s ein positiver Wert sein. (Die Möglichkeit, daß s unendlich ist, kommt auch nicht in Betracht; denn nach Hinzufügung von positiven Gliedern erhält man aus s die Reihe

$$\left(1 - \frac{1}{2}\right) + \left(\frac{1}{2} - \frac{1}{3}\right) + \left(\frac{1}{3} - \frac{1}{4}\right) + \left(\frac{1}{4} - \frac{1}{5}\right) + \left(\frac{1}{5} - \frac{1}{6}\right) + \dots,$$

und diese hat den endlichen Wert 1, weil sich beim Weglassen der Klammern je zwei aufeinanderfolgende entgegengesetzt gleiche Glieder aufheben.)

Widersprüche dieser Art erwiesen es als unumgänglich notwendig, das Grundsätzliche an der Infinitesimalrechnung zu untersuchen, um auf diese Weise zu einer sicheren Handhabung der Methode zu gelangen und feste Regeln für das mathematische Schließen zu gewinnen. Die Inifinitesimalrechnung mußte auf eine sichere logische Grundlage gestellt werden. Diese Aufgabe bereitete zunächst große Schwierigkeiten. Die ersten Versuche einer strengen Begründung der Infinitesimalrechnung schlugen fehl, und es war lediglich den ganz offenkundigen großartigen Erfolgen der Infinitesimalrechnung zu verdanken, daß man an der Möglichkeit einer logisch einwandfreien Grundlegung nicht irre wurde.

Heute ist es uns ganz geläufig, wie sich die Schlußfolgerungen der Infinitesimalrechnung auf strenge Art ausführen lassen. Um diese sichere Methode des Schließens zu finden, war vor allem die Einsicht nötig, daß die gewöhnlichen Rechenregeln nicht ohne weiteres auf unendliche Prozesse übertragen werden dürfen. Das bedeutete zu jener Zeit eine große Entdeckung; denn die damaligen Mathematiker hielten es für selbstverständlich, daß man mit unendlichen Summen, Produkten usw. genauso wie mit endlichen rechnen könne.

Diese Voraussetzung ist auch bei den drei angeführten Beispielen die Ursache für das Zustandekommen des Widerspruchs. In den beiden ersten Fällen wird ohne weiteres vorausgesetzt, daß eine unendliche Reihe immer einen bestimmten Wert hat, und in dem dritten Falle nimmt man als selbstverständlich an, daß der Wert einer unendlichen Reihe bei einer Umordnung der Glieder nicht geändert wird.

Indem man sich nun des Näheren die Bedeutung unendlicher Prozesse überlegte, entstanden alle jene uns jetzt ganz vertrauten Begriffsbildungen: die Begriffe der Konvergenz und Divergenz, die Unterscheidung von gleichmäßiger und ungleichmäßiger, von bedingter und unbedingter Konvergenz; man beachtete ferner die Reihenfolge der Grenzübergänge und den Unterschied zwischen der Annäherung an einen Wert und dem Annehmen des Wertes.

Ein Beispiel für die Wichtigkeit solcher Unterscheidungen bildet die Frage nach der Existenz eines Minimums von einer Zahlenmenge. Man sagt von einer Zahlenmenge (einer Gesamtheit von Zahlen), daß sie ein Minimum besitzt, wenn es eine Zahl der Menge gibt, die nicht größer ist als irgendeine der übrigen. Bei den Zahlenmengen, die nur aus endlich vielen Zahlen bestehen, ist die Existenz eines Minimums selbstverständlich. Zum Beispiel muß es in einer Gesamtheit von endlich vielen Menschen mindestens einen jüngsten geben (d. h. einen, welcher keinen der übrigen an Alter übertrifft). Dagegen braucht es in einer Gesamtheit von unendlich vielen Größen durchaus nicht immer eine kleinste zu geben. Das lehrt z. B. die Gesamtheit der unendlich vielen Zahlen

$$\frac{1}{1}, \frac{1}{2}, \frac{1}{3}, \frac{1}{4}, \frac{1}{5}, \ldots,$$

unter denen es ersichtlich keine kleinste gibt.

Auch geometrisch läßt es sich gut klar machen. Betrachten wir etwa die Gesamtheit derjenigen Linien, welche zwei gegebene Punkte A, B miteinander verbinden, und welche überdies die Eigenschaft haben, daß ihre Anfangsrichtung im Punkte A senkrecht ist zu der Geraden, welche durch A und B geht. Unter diesen Linien gibt es keine kürzeste; denn zu jeder von ihnen läßt sich eine andere angeben, die sich in ihrer Länge noch weniger als jene von der geraden Verbindungsstrecke AB unterscheidet. (Die Strecke AB selbst gehört nicht mit zu der betrachteten Gesamtheit.)

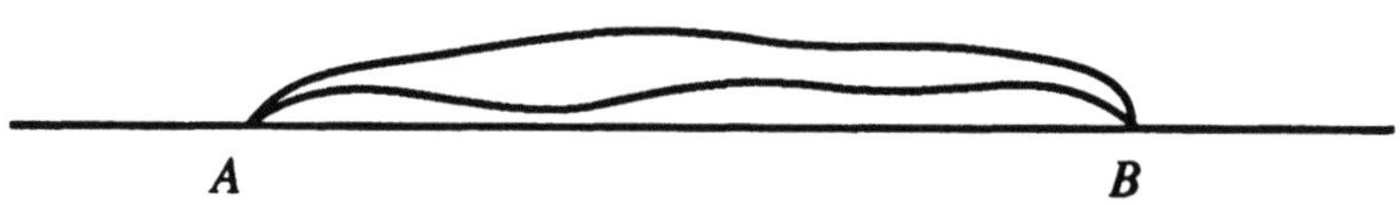

Fig. 4

Obwohl nun dieser Sachverhalt so einleuchtend ist, so ist man doch immer darüber hinweggegangen.

Als ein klassisches Beispiel hierfür ist ***Riemanns Theorie der Abelschen Funktionen*** anzuführen. Dieses Hauptwerk des großen Mathematikers Riemann ist auf einen Fehlschluß aufgebaut, indem ohne weiteres angenommen wird, daß unter gewissen Integralen, die in unendlicher Zahl vorhanden sind, eines vom kleinsten Werte existiert. Erst Weierstraß hat auf diesen Irrtum aufmerksam gemacht, und es bedurfte dann noch einer Reihe von schwierigen Untersuchungen, bis es gelang, die Grundlage der Riemannschen Theorie durch den Beweis seiner Grundannahme sicherzustellen.* Das methodisch Anfechtbare bei diesem Vorgehen Riemanns besteht darin, daß er unbewußt jene Voraussetzung der Existenz eines Minimums gemacht hat. Hätte er diese Annahme mit Bewußtsein als Axiom an die Spitze seiner Theorie gestellt, so wäre an diesem Verfahren nichts auszusetzen. Er hätte dann allerdings ein überflüssiges Axiom eingeführt, aber ein methodischer Fehler läge darin nicht.

Es ist vielmehr ein durchaus zulässiges wissenschaftliches Verfahren, daß man einen Satz, der plausibel erscheint, für den man aber keinen Beweis hat, vorläufig als Postulat hinstellt und zusieht, was sich unter seiner Annahme für Folgerungen ergeben. Dies kommt darauf hinaus, daß man den betreffenden Satz zu den Axiomen hinzunimmt. Gelingt es nachher, den Satz zu beweisen, so wird damit das Axiom als überflüssig erwiesen.

Dieses Verfahren der Einführung eines Postulates hat in der Mathematik mehrfach Anwendung gefunden, insbesondere auch in einem Gebiete, wo man es am wenigsten erwarten sollte, nämlich in der Zahlentheorie. Bekannt ist hier ein Postulat, welches von Legendre aufgestellt wurde. Es handelt sich dabei um Folgendes. Sind m und n zwei ganze Zahlen, so heißt die Zahl m quadratischer Rest von n, wenn sie sich in der Form

$$(a^2 + bn)$$

darstellen läßt (wobei a und b wiederum ganze Zahlen bedeuten). Es besteht nun bei zwei Primzahlen p, q, von denen mindestens eine die Form $(4k + 1)$ hat, immer die Beziehung, daß p quadratischer Rest von q ist, falls q quadratischer Rest von p ist. Diesen Satz, welcher als "quadratisches Reziprozitätsgesetz" bezeichnet wird, wollte Legendre beweisen. Das ist ihm nicht geglückt; aber er konnte doch so viel zeigen, daß der Satz sich als Folgerung ergibt, wenn man die plausible Annahme zugrundelegt, daß unter den Zahlen einer arithmetischen Reihe

$$a,\ a + d,\ a + 2d,\ a + 3d,\ \ldots,$$

* Die Resultate der Riemannschen Theorie waren inzwischen auf anderem Wege bewiesen worden.

worin a und d zwei teilerfremde ganze Zahlen sind, stets unendlich viele Primzahlen vorkommen. Er bewies also das Reziprozitätsgesetz auf Grund des Postulates über die arithmetischen Reihen. Das war ein ganz korrektes Vorgehen. Aber natürlich wäre es unbefriedigend gewesen, dabei stehenzubleiben. Man mußte nach einem Beweise für jenes Postulat suchen. Und in der Tat ist ja, wie Sie wissen, dieser Beweis erbracht worden, nachdem schon vorher Gauss das Reziprozitätsgesetz auf direktem Wege bewiesen hatte.

Ein anderes berühmtes Beispiel dieser Art bildet ein von Riemann in die *analytische Zahlentheorie* eingeführtes Postulat. Riemann zeigte, wie die Verteilung der Primzahlen mit den Eigenschaften einer gewissen transzendenten Funktion $\zeta(s)$ der komplexen Variablen s zusammenhängt. Seine Untersuchung über die Primzahlen gipfelt darin, daß er eine sehr elegante und scharfe Abschätzung für die Anzahl der Primzahlen unter einer gegebenen Grenze ableitet. Diese Ableitung gilt jedoch nur unter der Bedingung, daß eine gewisse Annahme über die Funktion $\zeta(s)$ zutrifft, daß nämlich eine mit der Funktion $\zeta(s)$ nahe verwandte Funktion nur reelle Nullstellen besitzt. Diese Annahme wird von Riemann ausdrücklich als unbewiesene Voraussetzung hervorgehoben. Auch heute ist man noch nicht imstande, sie zu beweisen. Man hat aber allen Grund zu vermuten, daß sie richtig ist, da man ihre Konsequenzen nach verschiedenen Richtungen sehr eingehend verfolgt hat, ohne daß sich je ein Widerspruch ergeben hätte.

Die Überwindung der Schwierigkeiten in der Grundlegung der Infinitesimalrechnung geschah hauptsächlich durch Cauchy und Weierstraß, denen es gelang, die Methode der Grenzprozesse begrifflich scharf zu fassen und auf feste Regeln zu bringen, so daß alle Unklarheit und Unsicherheit beseitigt wurde. Seitdem ist ein neuer Geist in die mathematische Analysis eingezogen. Eine sichere Bewegungsfreiheit ist erreicht: In den Schlußweisen herrscht eine völlig befriedigende Strenge, und Widersprüche sind hier nicht mehr zu befürchten.

Nun werden wir aber bei der weiteren Entwicklung der Mathematik von neuem gewahr, daß das mathematische Schließen nicht untrüglich ist. Es war gerade die volle Herrschaft über die Infinitesimalrechnung, welche den Mathematikern das Gefühl der Sicherheit gab und sie zu neuen kühnen Unternehmungen ermutigte. Diese neue Entwicklung ging von Cantor aus, dessen Gedanken an Originalität in der mathematischen Wissenschaft ihresgleichen suchen und an grundsätzlicher Bedeutung der Infinitesimalrechnung kaum nachstehen. Durch ihn wurde ein ganz neues Wissensgebiet, die *Mengenlehre*, geschaffen. Über die Anfangsgründe dieser Wissenschaft will ich hier einige Ausführungen geben.

Cantor machte zunächst die Entdeckung, daß die rationalen Zahlen sich zählen (numerieren) lassen, daß man also jede rationale Zahl mit einer Nummer versehen kann in der Weise, daß verschiedene rationale Zahlen auch verschiedene Nummern haben. Dies erscheint auf den ersten Blick geradezu unbegreiflich, da ja die rationalen Zahlen innerhalb des Kontinuums der reellen Zahlen überall dicht liegen, nämlich

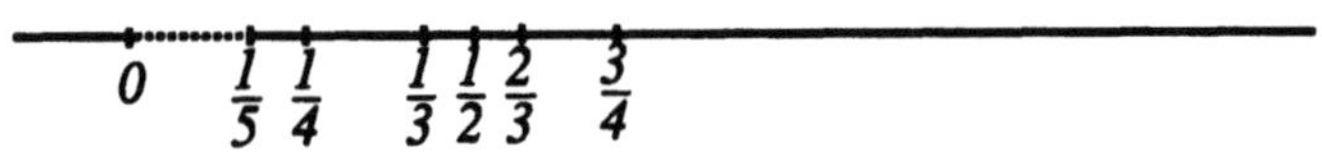

Fig. 5

also jeder reelle Punkt beliebig eng durch rationale eingeschlossen wird,—ja schon die gewöhnliche Dezimalbruchentwicklung einer reellen Zahl beruht ja darauf, daß sie schon durch die speziellen rationalen Zahlen mit den Nennern 10, 100, 1000 beliebig gut angenähert werden kann. Daß trotzdem die Abzählung möglich ist, kann man auf folgende Weise einsehen.* Man legt ein quadratisches Schema an, das nach rechts und nach unten offen ist.

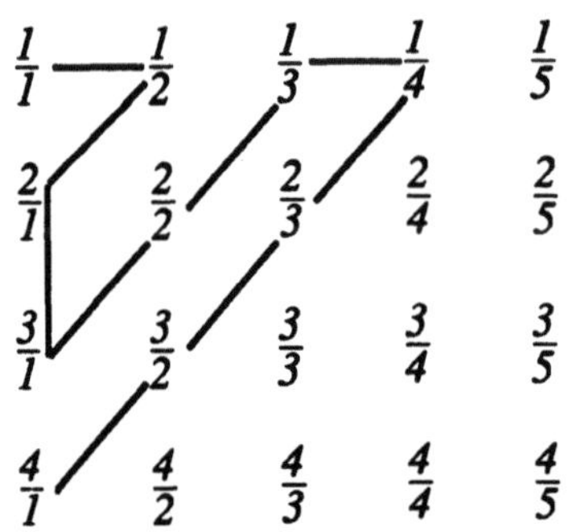

Fig. 6

Hierin weist man dem Bruche m/n dasjenige Feld zu, welches in der m-ten Zeile und in der n-ten Spalte liegt. Auf diese Weise werden alle rationalen Brüche in das Schema eingeordnet. Nun kann man (wie in der Figur angedeutet ist) eine Linie ziehen, welche durch alle Felder des Schemas nacheinander hindurchgeht. Diese Linie verfolge man und schreibe die Brüche in der Reihenfolge auf, in der die zugehörigen Felder von der Linie

* Der Einfachheit halber wollen wir die Betrachtung auf die *positiven* rationalen Zahlen beschränken.

getroffen werden. Läßt man nun noch alle diejenigen Brüche weg, die mit einem vorhergehenden dem Werte nach übereinstimmen, dann hat man die rationalen Zahlen in eine Reihe geordnet:

$$\frac{1}{1}, \frac{1}{2}, \frac{2}{1}, \frac{3}{1}, \frac{1}{3}, \frac{1}{4}, \frac{2}{3}, \frac{3}{2}, \frac{4}{1}, \frac{5}{1}, \ldots,$$

und indem man jeder rationalen Zahl ihre Stellenzahl in dieser Reihe als Nummer zuweist, ist die Numerierung der rationalen Zahlen vollzogen.

Den hiermit bewiesenen Satz pflegt man so zu formulieren: Die Menge der rationalen Zahlen ist abzählbar. Man nennt nämlich allgemein eine Menge abzählbar, wenn die zu ihr gehörigen Gegenstände (die "Elemente" der Menge) sich numerieren lassen. Allgemeiner können wir jenen Satz so aussprechen: Die Punkte eines Kontinuums erster Art (wie wir es nannten) bilden immer eine abzählbare Menge. Denn ein Kontinuum erster Art läßt sich ja stets durch das System der rationalen Zahlen repräsentieren.

Das erhaltene Ergebnis legt nun die Frage nahe, ob es denn überhaupt Mengen gibt, die nicht abzählbar sind. Die Entscheidung brachte Cantors zweite Entdeckung, daß die Menge der reellen Zahlen jedenfalls nicht abzählbar ist. Der Beweis hiervon wird indirekt geführt. Angenommen, die Menge aller reellen Zahlen wäre abzählbar, so müßte dasselbe auch für jede Teilmenge gelten. Denn durch eine Numerierung aller reellen Zahlen würden ja auch die Zahlen der Teilmenge in eine bestimmte Reihenfolge gebracht. Insbesondere müßte es also auch eine Abzählung aller derjenigen reellen Zahlen im Intervall von 0 und 1 auftreten. Da diese Dezimalbruchentwicklungen den durch sie dargestellten reellen Zahlen umkehrbar eindeutig entsprechen, so müßten auch sie numeriert werden können, das heißt, es müßte eine Reihe von Dezimalbrüchen

$$\alpha_1, \alpha_2, \alpha_3, \alpha_4, \ldots$$

geben, welche alle solchen Dezimalbrüche enthielte, bei denen vor dem Komma 0 und an jeder Dezimalstelle entweder 0 oder 1 steht.* Dieser Folgerung widerspricht aber die Tatsache, daß zu jeder Reihe

$$\alpha_1, \alpha_2, \alpha_3, \alpha_4, \ldots,$$

die aus Dezimalbrüchen von der betrachteten Art besteht, ein Dezimalbruch α bestimmt werden kann, der wiederum von der speziellen Art ist (d. h. mit 0 vor dem Komma beginnt und in den Dezimalstellen nur die Ziffern 0 oder 1 enthält), der aber nicht mit in der Reihe vorkommt. Denn man hat

* Wir fassen alle Dezimalbrüche als unendliche auf, indem wir bei einer abbrechenden Entwicklung die Periode 0 hinzufügen.

ja bei jeder Dezimalstelle die Wahl zwischen 0 und 1 frei und kann daher so verfügen, daß die Ziffer der ersten Dezimalstelle verschieden ist von der bei α_1, die der zweiten Dezimalstelle verschieden von der bei α_2, allgemein die Ziffer der n-ten Dezimalstelle verschieden ist von der bei α_n. Auf diese Weise wird ein bestimmter Dezimalbruch α von der speziellen Art definiert, und dieser unterscheidet sich von jedem der Dezimalbrüche aus der Reihe $\alpha_1, \alpha_2, \alpha_3, \ldots$ in mindestens einer Stelle; er kommt also gewiß nicht in der Reihe vor. Somit führt die Annahme, daß die reellen Zahlen sich numerieren lassen, zu einem Widerspruch. Die Menge der reellen Zahlen ist also nicht abzählbar, oder anders ausgedrückt: Die Punkte eines Kontinuums zweiter Art lassen sich nicht abzählen. Während also die rationalen Zahlen sich umkehrbar eindeutig den ganzen Zahlen zuordnen lassen, ist dies für die reellen Zahlen nicht mehr möglich. Es kann daher auch zwischen den Punkten eines Kontinuums erster Art und denen eines Kontinuums zweiter Art keine eindeutige Zuordnung hergestellt werden.

Der fundamentale Unterschied, der sich hier zwischen der Menge der rationalen Zahlen und der Menge der reellen Zahlen (zwischen dem Kontinuum erster Art und dem zweiter Art) zeigt, ist ganz von derselben Art, wie er zwischen zwei endlichen Mengen stattfindet, wenn die Anzahl ihrer Elemente verschieden ist. Denn daß zwei endliche Mengen aus einer ungleichen Anzahl von Gegenständen bestehen, erkennt man ja daran, daß sich die Gegenstände der einen Menge denen der anderen Menge nicht umkehrbar eindeutig zuordnen lassen.

So wurde Cantor durch seine Entdeckungen dazu geführt, den Begriff der Anzahl für unendliche Mengen zu verallgemeinern. Und zwar nannte er dasjenige, was bei einer endlichen Menge die Anzahl der Elemente heißt, allgemein die "Mächtigkeit" einer Menge. Jeder Menge wird also nach Cantor eine bestimmte Mächtigkeit zugeschrieben. Zwei Mengen M, N heißen von gleicher Mächtigkeit, wenn sich die Elemente von M denen von N umkehrbar eindeutig zuordnen lassen, N heißt von größerer Mächtigkeit als von M, wenn die Elemente von M zwar nicht den sämtlichen Elementen von N, aber einem Teil davon umkehrbar eindeutig zugeordnet werden können.

Die kleinstmögliche Mächtigkeit einer unendlichen Menge ist diejenige Mächtigkeit $\aleph_0$, welche der Menge der ganzen Zahlen zukommt. Gemäß der Definition der Abzählbarkeit hat eine jede abzählbare Menge, insbesondere also die Menge der rationalen Zahlen, diese Mächtigkeit $\aleph_0$. Der Satz von Cantor über die reellen Zahlen besagt nun: Es gibt Mengen, die von größerer Mächtigkeit sind als $\aleph_0$, nämlich die Menge der reellen Zahlen ist eine solche Menge.

Nachdem diese Erkenntnis gewonnen war, entstand naturgemäß die Frage, ob etwa zwischen der Mächtigkeit $\aleph_0$ und der Mächtigkeit des Kontinuums zweiter Art noch weitere Mächtigkeiten vorhanden sind (ob es also nicht eine Teilmenge des Kontinuums zweiter Art gibt, die weder abzählbar, noch auch auf die Menge der reellen Zahlen umkehrbar eindeutig abbildbar ist), oder ob die Mächtigkeit des Kontinuums zweiter Art die nächste Mächtigkeit darstellt.

Wenngleich nun Cantor dies nicht zu entscheiden vermochte, so ist es ihm doch gelungen, so viel sicher zu stellen, daß es eine unmittelbar auf $\aleph_0$ folgende Mächtigkeit ganz gewiß gibt. Ich möchte die Gedankengänge, durch welche Cantor zu diesem Nachweis gelangt ist, wegen ihrer grundsätzlichen Wichtigkeit etwas näher andeuten.

Durch die Unterscheidung der Mächtigkeiten haben wir *Kardinalzahlen* (Anzahlen im weiteren Sinn) für unendliche Mengen eingeführt. Nun werden wir auch versuchen, den Begriff der *Ordnungszahl* auf unendliche Mengen zu übertragen. Hierbei kommt es zunächst darauf an, den Begriff der Reihenfolge so weit zu fassen, daß er auch auf nicht abzählbare Mengen Anwendung finden kann. Dieser Anforderung entspricht der Begriff der *Wohlordnung.* Man nennt eine Menge wohlgeordnet, wenn eine Rangordnung zwischen den Elementen festgesetzt ist, bei welcher ein Element das erste ist und zu jedem Abschnitte der Menge ein nächstfolgendes Element existiert.*

Es macht sich nun in Hinsicht auf die Wohlordnung ein charakteristischer Unterschied zwischen endlichen und unendlichen Mengen geltend. Bei einer endlichen Menge ist die Wohlordnung gleichbedeutend mit der gewöhnlichen Ordnung in einer Reihenfolge, d. h. mit einer Numerierung der Elemente. Eine solche ist im wesentlichen nur auf eine Art möglich; die verschiedenen Numerierungen einer endlichen Menge können alle ineinander durch Vertauschung der Elemente übergeführt werden, und die Gesamtheit der zu verwendenden Ordnungszahlen ist durch die Anzahl der Elemente eindeutig bestimmt. Ganz anders verhält es sich bei unendlichen Mengen. Bei diesen kann die Wohlordnung stets auf (unendlich viele) verschiedene Arten ausgeführt werden. Nehmen wir das einfachste Beispiel, die Menge der positiven ganzen Zahlen. Für diese gibt es zunächst die natürliche Reihenfolge:

$$1\,2\,3\,4\ldots,$$

und aus dieser werden durch Vertauschungen der Zahlen die verschiedenen möglichen Abzählungen der ganzen Zahlen erhalten. Hiermit sind aber noch keinewegs alle Wohlordnungen erschöpft. Zum Beispiel gibt es die Wohlordnung

$$2\,3\,4\,5\,6\ldots 1,$$

* Eine Anordnung zwischen den Elementen bedeutet, daß von je zwei Elementen eines als das frühere und eines als das spätere bestimmt ist; und unter einem Abschnitt einer (mit einer Rangordnung versehenen) Menge M versteht man eine solche echte Teilmenge von M, in welcher zugleich mit jedem ihr angehörenden Element a von M auch jedes der Rangordnung nach frühere Element von M enthalten ist.

in welcher die Eins auf alle übrigen Zahlen folgt. Diese kann offenbar nicht durch eine Vertauschung in die natürliche Reihenfolge übergeführt werden. Denn bei dieser folgt auf jede Zahl eine nächste, und jeder Zahl mit Ausnahme der ersten geht eine andere Zahl unmittelbar vorher, während hier der Eins keine Zahl nachfolgt und keine unmittelbar vorhergeht. Ebenso wie die Eins können wir auch mehrere Zahlen, z. B. 1, 2, 3 auf die übrigen Zahlen folgen lassen:

$$456\ldots 123.$$

Wir können ferner zwei Abzählungen vom Typus der natürlichen Zahlenreihe hintereinander setzen, indem wir z. B. auf die der Größe nach geordneten geraden Zahlen die ungeraden Zahlen (wiederum der Größe nach) folgen lassen:

$$2468\ldots 1357\ldots \quad .$$

Auch unendlich viele Abzählungen können aneinander gereiht werden, z. B. wenn zuerst die ungeraden Zahlen, dann die Vierfachen, dann die Achtfachen usw.

So ergibt sich eine große Mannigfaltigkeit von wesentlich verschiedenen Typen der Wohlordnung. Das Kennzeichnende dabei ist, daß wir zur Numerierung der Elemente nicht mehr mit den endlichen Ordnungszahlen auskommen. Schon bei der Wohlordnung

$$23456\ldots 1$$

können wir der Eins keine endliche Ordnungszahl zuweisen; denn die endlichen Nummern sind für die vorhergehenden Zahlen verbraucht. Wir müssen also über das Unendliche hinauszählen, indem wir "transfinite" Ordnungszahlen einführen. Diese transfiniten Ordnungszahlen bilden zugleich die Symbole zur Unterscheidung der verschiedenen Typen der Wohlordnung, in entsprechender Weise, wie die endlichen Ordnungszahlen zugleich als Bezeichnungen für die einzelnen Zahlen dienen (welche ja bei endlichen Mengen den Typus der Wohlordnung eindeutig bestimmen).

Man bezeichnet den Typus der natürlichen Zahlenreihe mit ω; die Typen der Wohlordnungen

$$234\ldots 1$$
$$3456\ldots 12$$
$$4567\ldots 123 \qquad \text{usw.}$$

werden bezüglich mit $\omega+1, \omega+2, \omega+3, \ldots$, der Typus

$$2468\ldots 13579\ldots$$

mit $\omega + \omega$ oder 2ω bezeichnet. Durch diese neue Methode des Zählens erhält man eine unbegrenzte Reihe von transfiniten Ordnungszahlen, deren Anfang durch folgendes Schema darstellt wird:

$$\begin{array}{l} 1, 2, 3, 4, 5, 6, \dots, \omega \\ \omega + 1, \omega + 2, \omega + 3, \dots, 2\omega \\ 2\omega + 1, 2\omega + 2, 2\omega + 3, \dots, 3\omega \\ \dots, 4\omega, \dots, 5\omega, \dots, 6\omega, \dots, \omega^2 \\ \dots, \omega^3, \dots, \omega^4, \dots, \omega^5, \dots, \omega^\omega, \dots \end{array}$$

Dieses Zählverfahren, welches sehr formal und abstrakt erscheint, können wir uns dadurch leichter verständlich machen, daß wir ihm eine geometrische Deutung geben. Es sei nämlich $f(x)$ eine Funktion, die für die Argumentwerte im Intervall von 0 bis 1 (einschließlich Null) definiert ist und der Bedingung

$$x < f(x) < 1$$

genügt. Mit Hilfe einer solchen Funktion, durch welche die Strecke von 0 bis 1 auf sich abgebildet wird, kann ich das allgemeinere Verfahren des Zählens folgendermaßen darstellen. Ich beginne mit dem Punkt $x_1 = 0$. Von diesem gehe ich nacheinander zu den Punkten

$$\begin{aligned} x_2 &= f(x_1) = f(0) \\ x_3 &= f(x_2) = f(f(0)) \\ x_4 &= f(x_3) = f(f(f(0))) \quad \text{usw.} \end{aligned}$$

über, von denen jeder folgende rechts von dem vorhergehenden liegt. Die Punkte $x_1, x_2, x_3, \dots$ nähern sich nun einem Grenzpunkt. Diesen nenne ich x_ω. Bei geeigneter Wahl von $f(x)$ wird nun x_ω noch kleiner sein als 1; und ich kann dann das Verfahren weiterführen, indem ich zu den Punkten

$$\begin{aligned} x_{\omega+1} &= f(x_\omega) \\ x_{\omega+2} &= f(x_{\omega+1}) = f(f(x_\omega)) \quad \text{usw.} \end{aligned}$$

fortschreite. Die Punktfolge $x_{\omega+1}, x_{\omega+2}, \dots$ hat wiederum einen Grenzpunkt, welchen ich $x_{2\omega}$ nenne. Ist $x_{2\omega}$ gleich 1, so bricht das Verfahren ab, anderenfalls kann es weiter in der gleichen Weise fortgesetzt werden.

Auf diese Art erhalten wir eine anschauliche Darstellung für das transfinite Zählen.* Cantor hat nun bewiesen, daß die Menge aller der Ordnungszahlen, welche man durch die neue Methode des Zählens erhält, oder—was auf dasselbe hinauskommt—die Menge aller verschiedenen Typen von Wohlordnungen der ganzen Zahlen nicht mehr abzählbar ist, daß vielmehr ihre Mächtigkeit die nächste Kardinalzahl nach $\aleph_0$ bildet. Seine Vermutung geht dahin, daß also mit Hilfe jener Ordnungszahlen die Punkte des Kontinuums zweiter Art sich (in dem erweiterten Sinne) zählen lassen.

Dies sind die Gedanken, deren genaue Ausführung und weitere Verfolgung den Hauptinhalt der Mengenlehre ausmacht. Mit dieser Wissenschaft der Mengenlehre ist es nun ähnlich gegangen wie mit der Infinitesimalrechnung. Sie entwickelte sich zunächst sehr erfolgreich. Trotzdem erstanden ihr heftige Widersacher, welche die neuen Methoden mit dem Geiste der mathematischen Strenge nicht für vereinbar hielten. Insbesondere war es Kronecker, der mit allen Mitteln gegen die Mengenlehre auftrat. Zugunsten Cantors sprach jedoch die große Fruchtbarkeit seiner Theorie. Es kam aber dann auch hier wieder so, daß sich bei der Anwendung der neuen Schlußweisen Widersprüche herausstellten.

Ein berühmtes Beispiel einer mengentheoretischen Schlußweise, die sich als fehlerhaft erwiesen hat, findet sich in Dedekinds Schrift "Was sind und was sollen die Zahlen?", welche eine der bedeutendsten Untersuchungen über die Grundlagen der Arithmetik bildet. Dedekind stellt sich hier die Aufgabe, eine strenge Begründung zu geben für die (insbesondere in der Zahlentheorie) grundlegende mathematische Schlußweise, welche man als den Schluß von n auf $(n+1)$ oder auch als das Prinzip der vollständigen Induktion bezeichnet. Diese würde ihm auch gelungen sein, wenn sich nicht im Anfang seiner Beweisführung eine fehlerhafte Argumentation fände. Es verhält sich damit ähnlich wie mit dem früher erwähnten Riemannschen Fehlschluß: Dedekind braucht zu seinen Überlegungen einen Satz als notwendige Grundlage, nämlich den Satz, daß es unendliche Mengen gibt. Hätte er diesen als Axiom seiner Untersuchung zugrundegelegt, so wäre sein Verfahren völlig korrekt; allerdings hätte er dann eine prinzipielle Frage unentschieden gelassen. Nun sucht Dedekind aber diesen Satz zu beweisen, und dazu verfährt er folgendermaßen.

Er gibt zunächst die Definition der unendlichen Menge: eine Menge heißt unendlich, wenn es möglich ist, die ganze Menge auf einen echten Teil

---

* Allerdings muß das Verfahren stets an einer Stelle abbrechen, in dem der Punkt 1 schließlich einmal erreicht wird. Andererseits kann aber für jede Ordnungszahl α, zu der man bei der allgemeineren Methode des Zählens gelangt, eine Funktion $f(x)$ gefunden werden, bei welcher das Verfahren der anschaulichen Darstellung jedenfalls noch bis zu einem dieser Ordnungszahl entsprechenden Punkte x_α führt.

umkehrbar eindeutig abzubilden. (Daß dies die charakteristische Eigenschaft für eine unendliche Menge ist, habe ich ja bereits früher hervorgehoben.) Nun betrachtet er die Menge A aller Dinge, welche Gegenstände des Denkens sein können. Um von dieser zu zeugen, daß auf sie die Definition der unendlichen Menge zutrifft, ordnet er jedem Element a von A, also jedem möglichen Gegenstande des Denkens den Gedanken a' an diesen Gegenstand zu. Da jeder solche Gedanke wiederum Gegenstand des Denkens sein kann, so bilden die Dinge eine Teilmenge A' von A, und zwar ist A' eine echte Teilmenge von A, denn in der Menge A' kommen ja nur Gedanken vor, und es gibt jedenfalls Gegenstände des Denkens, welche selbst keine Gedanken sind, z. B. das eigene Ich. Es wird also durch die Beziehung zwischen a und a' die Menge A auf eine echte Teilmenge abgebildet. Und diese Abbildung ist umkehrbar eindeutig; denn durch den Gegenstand a ist der Gedanke a' an den Gegenstand a und umgekehrt auch durch diesen Gedanken a' der Gegenstand a (als dasjenige, worauf sich der Gedanke bezieht) in eindeutiger Weise bestimmt. Die Menge A ist demnach (im Sinne der Definition) eine unendliche Menge, und die Existenz unendlicher Mengen scheint somit erwiesen zu sein.

Nun hat sich aber diese Schlußweise als durchaus unzulässig herausgestellt. "Die Menge alles Denkbaren", das ist gerade ein typisches Beispiel einer unerlaubten Begriffsbildung, die wir unter keinen Umständen anwenden dürfen, wenn wir nicht in Widersprüche geraten wollen. Die prägnanteste Form eines derartigen Widerspruches ist das ***Russell-Zermelosche Paradoxon.*** Hierbei handelt es sich nicht um die Menge alles Denkbaren überhaupt, sondern speziell um die Menge aller denkbaren Mengen. Daß bereits diese Zusammenfassung unstatthaft ist, zeigt sich folgendermaßen.

Man kann die Mengen danach einteilen, ob unter ihren Elementen die Menge selbst vorkommt oder nicht. Hierbei kann es dahingestellt bleiben, ob es überhaupt möglich ist, daß eine Menge unter ihren eigenen Elementen vorkommt. Jedenfalls aber gibt es solche Mengen, die sich nicht selbst als Element enthalten. Zum Beispiel kann die Menge aller Menschen sich nicht selbst enthalten, da diese Menge doch nicht selbst ein Mensch ist. Es sei nun $N = \{N_1, N_2, \ldots\}$ die Menge aller der Mengen N_i, die sich selbst als Element nicht enthalten, d. h., die nicht selbst unter ihren eigenen Elementen vorkommt. Dann sind nur zwei Fälle möglich: Entweder gehört N zu den Mengen, die sich selbst nicht enthalten, oder N gehört zu den Mengen, die sich selbst enthalten. Beide Annahmen führen jedoch auf einen Widerspruch. Denn im ersten Falle müßte gemäß der Definition von N die Menge N unter den Elementen von N vorkommen, während die Annahme gerade dies ausschließt, im zweiten Falle würde die Menge N als Element eine Menge (nämlich N) enthalten, welche sich selbst enthält, entgegen der Definition von N. Dieser Widerspruch läßt sich nur dadurch lösen, daß die Begriffsbildung der Menge aller Mengen als logisch unzulässig aufgegeben wird.

Man könnte nun denken, daß das Fehlerhafte der Zusammenfassung aller Mengen darauf beruht, daß der Allgemeinbegriff der Menge zu vage ist. Dieser Umstand ist jedoch nicht wesentlich, vielmehr gibt es auch im Gebiete des rein Mathematischen derartige Mengenbildungen, die zunächst ganz unverfänglich erscheinen, von denen sich aber dann zeigt, daß sie auf Widersprüche führen. Ein Beispiel hierfür ist die Menge aller Mächtigkeiten, deren Einführung auch zu Widersprüchen Anlaß gibt. Durch die aufgedeckten Widersprüche wurde nun das ganze Gebäude der Mengenlehre gefährdet. Es braucht ja eine Theorie nur einen einzigen Widerspruch zu enthalten, dann wird alles hinfällig, indem dann jede Behauptung sowohl bewiesen wie auch widerlegt werden kann. Man mußte also die mengentheoretischen Schlußweisen schärfer präzisieren. Dies ist in dem Sinne gelungen, daß man Axiomsysteme aufgestellt hat, aus denen alle wertvollen Ergebnisse der Mengenlehre abgeleitet werden können, während andererseits diejenigen Schlußweisen, welche vordem zu den Widersprüchen führten, dadurch ausgeschlossen werden.

Nun geht aber die Problemstellung noch weiter; denn die Prinzipien der Mengenlehre stehen in so engem Zusammenhang mit den allgemeinen Grundlagen des mathematischen Schließens, daß eine befriedigende Grundlegung der Mengenlehre nur möglich ist, im Rahmen einer allgemeinen Systematisierung der Methoden des mathematischen Beweises überhaupt. So wird die Mathematik dazu gedrängt, den mathematischen Beweis selbst zum Gegenstand der Untersuchung zu machen, ähnlich wie etwa der Astronom den Apparat untersuchen muß, mit dem er seine Messungen ausführt. Hieran zeigt sich besonders deutlich, wie stark die landläufige Ansicht über das mathematische Schließen von dem wirklichen Sachverhalt abweicht: Die Methode des mathematischen Schließens ist so wenig selbstverständlich, daß ihre genaue Festlegung vielmehr noch als eine schwierige Aufgabe aussteht.

Es ist auch keine Aussicht vorhanden, daß sich die mathematischen Schlußweisen vollständig auf die Schlüsse der üblichen Logik zurückführen lassen. Man wird sich vielmehr damit begnügen müssen, ein System von Grundsätzen und Regeln aufzustellen, aus denen einerseits die gesamte Mathematik entwickelt werden kann, ohne daß dem gewohnten mathematischen Schließen irgendwie unnötig Gewalt angetan wird, und von denen sich andererseits erkennen läßt, daß sie niemals zu Widersprüchen führen können.

Ansätze hierzu sind in der mathematischen Logik bereits gemacht worden: Doch bleibt das Meiste erst noch zu leisten. Zum Beispiel ist noch nicht einmal wirklich bewiesen worden, daß die Regeln für das gewöhnliche mathematische Rechnen stets zu übereinstimmenden Ergebnissen führen müssen. Noch viel weiter sind wir davon entfernt, einen Beweis für die Widerspruchslosigkeit der Arithmetik zu besitzen; vielmehr wird diese überall da, wo man sonst die Widerspruchslosigkeit einer mathematischen Theorie beweist, immer schon vorausgesetzt. Auch noch andere, den mathematischen Beweis betreffende Fragen bedürfen der Klärung. Zum Beispiel ist man noch gänzlich im Ungewissen darüber, ob jedes arithmetische Problem (auf Grund der üblichen Axiome der Arithmetik) eine bestimmte Lösung hat. So knüpfen sich an das mathematische Schließen mannigfaltige und schwierige Fragen. Die Lösung dieser Probleme würde unsere Einsicht in das Wesen des logischen Denkens in erheblichem Maße fördern.

Hiermit möchte ich meine Ausführungen über das mathematische Denken abschließen. Überblickt man den Gang unserer Betrachtungen, so kann man sagen, daß die drei Gesichtspunkte, nach denen wir die primitive Auffassung über die Mathematik modifiziert haben, nach dem Muster des philosophischen Schemas: "Begriff, Urteil, Schluß" sich gruppieren. Wir stellten fest, daß die Begriffsbildungen der Mathematik nicht willkürlich sind, vielmehr sowohl aus äußeren wie auch aus inneren Gründen sich systematisch aufbauen. Sodann betrachteten wir die Grundurteile mathematischer Disziplinen und fanden, daß eine absolute Gewißheit der Ausgangssätze kein notwendiges Erfordernis zur erfolgreichen Anwendung der mathematischen Methode bildet. Und von dem mathematischen Schließen erkannten wir, daß es sich nicht auf die üblichen Methoden der Logik beschränkt und daß es auch nicht frei ist von Irrtümern und von Problemen. Nach ähnlichen Gesichtspunkten wollen wir nun auch die Physik untersuchen.

ZWEITER TEIL:

DIE LANDLÄUFIGE AUFFASSUNG VON DER PHYSIK UND IHRE BERICHTIGUNG

4. KAPITEL: PHYSIKALISCHE BEGRIFFSBILDUNGEN

Die landläufige Auffassung von der Physik läßt sich etwa folgendermaßen wiedergeben: Es gibt unabänderliche Naturgesetze, nach denen der Ablauf der Naturvorgänge stattfindet. Diese Naturgesetze müssen gefunden werden. Dazu werden verschiedene Hypothesen aufgestellt, und es entscheidet dann zwischen diesen das Experiment. Hat man die Gesetze gefunden, so sind damit die Naturvorgänge erklärt, und die Aufgabe der Physik ist dann (in theoretischer Hinsicht) gelöst.

Wir wollen uns nun klar machen, daß diese Ansicht unter verschiedenen Gesichtspunkten erheblich modifiziert werden muß. Das Erste, was wir daran zu beanstanden haben, ist folgendes: Ähnlich wie nach der primitiven Vorstellung von der Mathematik diese in einem bloßen Haufen logischer Schlüsse besteht, so würde nach der populären Auffassung von der Physik das Ergebnis der physikalischen Wissenschaft nichts anderes sein als eine bloße Ansammlung von Gesetzen, welche getrennt nebeneinander stehen. Diese Vorstellungsweise rührt davon her, daß man nur solche physikalischen Gesetze im Auge hat, die in einfacher Weise an die Wahrnehmung anknüpfen, von der Art wie z. B. das Hebelgesetz, die Stoßgesetze, der Satz von der geradlinigen Fortpflanzung des Lichtes, das Gesetz, wonach sich die Körper bei Erwärmung ausdehnen, der Satz, daß der Druck eines Gases umgekehrt proportional ist zu dem Volumen, das Gesetz der Ablenkung einer Magnetnadel durch den elektrischen Strom. Diese Gesetze haben alle das gemeinsam, daß man sie fast unmittelbar konstatieren kann. Sie enthalten nicht viel anderes als die Wiedergabe von Beobachtungen, und der Unterschied gegenüber einer gewöhnlichen Beschreibung eines Vorganges besteht nur darin, daß die Behauptung des Gesetzes allgemein und unter bestimmt formulierten Voraussetzungen ausgesprochen wird, daß größere Genauigkeit angewandt und von unwesentlichen Umständen abgesehen wird.

Auf dem anfänglichen Stadium der Physik haben die Gesetze alle diesen Charakter und sie stehen in der Tat unverbunden nebeneinander. Aber mit solch einer Sammlung von Gesetzen gibt sich die Wissenschaft nicht zufrieden. Jene direkt an Hand der Wahrnehmungen gewonnenen (phänomenologischen) Gesetze bilden nicht das Endziel, sondern vielmehr nur den notwendigen Ausgangspunkt, bei welchem erst eigentlich die theoretische Problemstellung einsetzt. Diese geht darauf aus, die sehr mannigfaltigen, direkt feststellbaren Zusammenhänge auf weniger Tatsachen zurückzuführen.

Es werden also die zunächst vorgefundenen Gesetzlichkeiten ihrerseits zum Gegenstand der Erklärung gemacht.

Dies führt dazu, begriffliche Systeme (Fachwerke von Begriffen) zu konstruieren. Es entwickeln sich Theorien, welche große Gebiete von Erscheinungen umfassen, wie die Mechanik, die Wärmelehre, die Theorie des elektromagnetischen Feldes. Die verschiedenen Theorien, zwischen denen zunächst kein Zusammenhang besteht, sucht man dann zu einem einzigen System zusammenzuschließen. Bei dieser fortschreitenden Systematisierung wird man dazu genötigt, sich in den Begriffsbildungen immer weiter von der unmittelbaren Beobachtung zu entfernen. Dies ist der allgemeine Gang der Entwicklung. Wir wollen nun diesen Prozeß genauer verfolgen. Dabei werden sich zwei große Entwicklungsstufen herausheben.

Die erste Wandlung in der physikalischen Naturbetrachtung, welche über den naiven Standpunkt hinausführt, beruht auf der Erwägung, daß die große Mannigfaltigkeit der Arten von physikalischen Erscheinungen, welche sich dem Versuch einer systematischen Zusammenfassung der verschiedenen durch Beobachtung gefundenen Gesetze zunächst als das wesentliche Hindernis entgegenstellt, in der Verschiedenheit unserer Sinne ihren Grund hat. Man sagt sich, daß die Unterscheidung der Phänomene des Lichtes, des Schalles, der Wärme usw. sich lediglich auf die Organe bezieht, mit denen wir die Naturvorgänge wahrnehmen, daß also eine wesentliche Verschiedenheit der physiklischen Vorgänge selbst dabei gar nicht vorzuliegen braucht.

Sehr frühzeitig hat man auch schon erkannt, daß der Schall durch Schwingungen der Luft erzeugt wird und daß die qualitativen Unterschiede der Töne sich vollkommen auf Unterschiede in der Art der Schwingungen zurückführen lassen, indem die Tonhöhe durch die Periodendauer, die Tonstärke durch den größten Ausschlag der Schwingung bestimmt wird und die Klangfarbe sich aus der Überlagerung verschiedener Schwingungen ergibt. Es zeigte sich dann auch von den Lichterscheinungen, daß sie auf Schwingungsvorgängen beruhen, wobei wiederum für die Qualität der Farbe die kinematischen Bestimmungen der Schwingungen maßgebend sind.

Gerade an den Farben läßt sich (vom Standpunkt unserer heutigen physikalischen Kenntnis) besonders deutlich nachweisen, wie willkürlich, objektiv betrachtet, diejenige Abgrenzung der Erscheinungen ist, welche sich nach der Art unserer Wahrnehmungen richtet. Die mit dem Auge wahrnehmbaren Lichtwellen umfassen nämlich der Frequenz nach kaum den Bereich einer Oktave. Und Schwingungen von ganz derselben Art sind es, die wir bei etwas kleinerer Schwingungszahl als strahlende Wärme wahrnehmen, die bei ganz langsamer Schwingung als Wellen der drahtlosen Telegraphie und bei den schnellsten Schwingungen als Röntgenstrahlen zur Geltung kommen.

Durch die Erkenntnis, daß das objektiv Physikalische bei den Erscheinungen der Töne und Farben in Bewegungsvorgängen besteht, wurde nun die Auffassung nahe gelegt, daß überhaupt alle Erscheinungen physikalisch durch

Bewegungen zu erklären sind, daß es also im Gegensatz zu der scheinbaren Vielgestaltigkeit der Naturvorgänge, welche uns durch die subjektiven Momente der Wahrnehmung vorgetäuscht wird, in Wirklichkeit nur eine Art von physikalischen Prozessen gibt, nämlich die Vorgänge der Bewegung. Diese Auffassung hat schon an sich betrachtet aus dem Grunde etwas sehr Plausibles, weil die Bewegungen etwas sind, was sich in der Vorstellung gut fassen und mit mathematischen Methoden behandeln läßt. Die möglichen Formen der Bewegung lassen sich ja mathematisch vollkommen überblicken, indem zu den geometrischen Beziehungen nur die zeitliche Veränderlichkeit hinzugenommen werden muß, so daß die Lehre von den möglichen Bewegungen, die Kinematik, sich als eine Theorie a priori entwickeln läßt.

Nun gewann aber die Ansicht, daß alle physikalischen Vorgänge als Bewegungen zu deuten sind, noch eine besonders starke Stütze dadurch, daß es Newton gelang, die verschiedenen Gesetze der sichtbaren körperlichen Bewegungen (Fallgesetz, Pendelgesetz, Hebelgesetz usw.) einschließlich der astronomischen Gesetze der Planetenbewegung aus ganz wenigen sehr einfachen Grundsätzen abzuleiten. Die Zusammenfassung von vorher ganz getrennten Tatsachen zu einem großzügigen, übersichtlichen theoretischen System der Mechanik wirkte so überzeugend, daß man mit diesem System glaubte, den Schlüssel zum Verständnis aller physikalischen Erscheinungen überhaupt gefunden zu haben.

So kam man zu der Forderung, alle physikalischen Tatsachen auf Mechanik zurückzuführen. Dieser methodische Leitgedanke erwies sich als äußerst fruchtbar und führte zu großen Erfolgen, die noch bis in die neueste Zeit hineinreichen. Es wurden die Unterschiede des festen, flüssigen und gasförmigen Zustandes auf bloße Unterschiede der Lagerung und Bewegung von Massenteilchen zurückgeführt, die Lehre vom Schall wurde vollkommen in die Mechanik eingeordnet, ferner gelang es, die ganze Thermodynamik aus der mechanischen Molekulartheorie der Gase unter Hinzunahme von Wahrscheinlichkeitsbegriffen abzuleiten, und diese neue Methode der Gastheorie erwies sich sogar gegenüber der früheren Thermodynamik bei der Erklärung mancher Erscheinungen als überlegen. Schließlich hat man auch auf die Chemie die mechanische Methode angewandt; durch Konstruktion von Atommodellen hat man die Verschiedenheit der chemischen Grundstoffe aus mechanischen Verhältnissen erklärt.

Bei allen diesen mechanischen Erklärungsweisen werden die Prinzipien der Mechanik durch die Annahme eingeführt, daß die Stoffe aus kleinen Teilchen (Molekülen) bestehen, deren Bewegungen den Gesetzen der festen Körper genügen. Es verbindet sich also mit dem Gedanken der mechanischen Deutung das Prinzip der Atomistik, und so bildet sich das *mechanistisch-atomistische Ideal* der Naturerklärung. Dieses ist schon von den alten griechischen Philosophen antizipiert worden, insbesondere von Demokrit, welcher lehrte, daß alle Sinnesqualitäten nur scheinbar existieren, und daß es in Wirklichkeit nichts anderes gebe als die Atome und den leeren Raum, in welchem sich die Atome bewegen ("ἄτομα καὶ κενὸν").

Das Ideal der vollkommen mechanistisch-atomistischen Erklärung der Naturerscheinungen kennzeichnet die *erste Stufe* in der Entwicklung der theoretischen Physik. In neuerer Zeit ist dieser Standpunkt am konsequentesten von Boltzmann in seiner Mechanik zur Geltung gebracht worden. Hier wird in der Tat alles zurückgeführt auf Atome und ihre Bewegungen. Die ganze Welt wird gedacht als bestehend aus einer endlichen Anzahl von Atomen. Die Atome sind Massenpunkte, das heißt Kraftzentren, umgeben mit einer Wirkungssphäre. Die Bewegungen der Atome werden durch die Kräfte bestimmt, welche die Atome aufeinander ausüben. Dies sind teils Anziehungs-, teils Abstoßungskräfte. (Auf der Anziehung beruht das Zusammenhalten, die Kohäsion der Materie, auf der Abstoßung die Undurchdringlichkeit der Materie.)

Danach ergibt sich folgender mathematischer Ansatz: die Atome werden numeriert und die Lage eines jeden Atoms wird durch seine rechtwinkligen Koordinaten

$$x_k,\ y_k,\ z_k \qquad (k = 1, 2 \ldots, n)$$

angegeben. Jedem Atom kommt eine Masse m_k zu. Die Bewegungen der Atome werden beschrieben, indem für jedes Atom die Lagekoordinaten als Funktionen der Zeit t:

$$x_k = \rho_k(t),\ y_k = \psi_k(t),\ z_k = \chi_k(t)$$

$$(k = 1, 2, 3, \ldots)$$

ausgedrückt werden. Und die Gesetze der Bewegung bestehen in einem System von gewöhnlichen Differentialgleichungen:

$$\frac{d^2 x_k}{dt^2} = F_k(x_1, y_1, z_1, x_2, \ldots, z_n, m_1, \ldots, m_n)$$

$$\frac{d^2 y_k}{dt^2} = G_k(x_1, y_1, z_1, x_2, \ldots, z_n, m_1, \ldots, m_n)$$

$$\frac{d^2 z_k}{dt^2} = H_k(x_1, y_1, z_1, x_2, \ldots, z_n, m_1, \ldots, m_n)$$

$$(k = 1, 2, \ldots, n),$$

wobei die Ausdrücke F_k, G_k, H_k, die Komponenten je eines Vektors bilden, der nur von den Massen und den gegenseitigen Entfernungen der Atome abhängt. Auf Grund dieser Differentialgleichungen müßte sich der gesamte

Verlauf des Naturgeschehens berechnen lassen, sofern für irgend einen Zeitpunkt die Massen, die Lagen und Geschwindigkeiten von allen Atomen bekannt wären.

So ist der Gedanke der Mechanik in der extremsten Form zum Ausdruck gebracht. Die große Mannigfaltigkeit der Erscheinungen hat man sich nach dieser Auffassung dadurch zu erklären, daß erstens die Konfigurationen, die sich aus den Lagenverhältnissen der Atome ergeben, sehr verschiedenartig sind, daß ferner wegen der ungeheuren Anzahl der Atome die Gesetze der großen Zahlen zur Geltung kommen und so gewisse typische Durchschnittserscheinungen verschiedener Art auftreten, welche scheinbar den Charakter von etwas Selbständigem haben, und daß überdies unsere Eindrücke von der Umwelt durch den sehr verwickelten Prozess der sinnlichen Wahrnehmung vermittelt sind, und infolge davon die Naturvorgänge sich uns nur bruchstückweise und sozusagen in einer komplizierten Projektion darstellen, bei der die bestehenden Zusammenhänge großenteils verdeckt werden.

Obwohl dieser Ansatz der Physik seiner Form nach höchst einfach ist, so erfordert natürlich doch die Ableitung der für unsere Erfahrung wichtigen Folgerungen aus den Grundgesetzen der Atombewegung einen großen Aufwand von mathematischen Überlegungen und Rechnungen. Zum Beispiel würde man schon zu der Erklärung des Vorhandenseins von festen Körpern eine lange Kette von Schlüssen gebrauchen. Das ist aber keine Eigentümlichkeit gerade der mechanischen Theorie, vielmehr ist die Anwendung umfangreicher mathematischer Hilfsmittel überhaupt unvermeidlich, wenn man einen vollständigen Einblick in alle physikalischen Zusammenhänge gewinnen will.

Durch die bedeutenden Erfolge, welche die Mechanistik erzielte, wurde man immer mehr in dem Glauben bestärkt, daß in der Tat das ganze physikalische Geschehen durch die Gesetze der Mechanik beherrscht werde. Es sind vor allem die Gebiete der Hydrodynamik und der Elastizitätstheorie, in welchen die mechanistische Erklärungsweise zu einer hohen Vollkommenheit der Systematik gelangt ist. Bei der mathematischen Entwicklung dieser Theorien spielt der Übergang zu *partiellen Differentialgleichungen* eine wesentliche Rolle. Diesen haben wir vom Standpunkt der mechanistisch-atomistischen Theorie als eine Art Mittelwertbildung aufzufassen, deren Anwendbarkeit darauf beruht, daß man sich nur für den Ablauf der Bewegungen im Großen interessiert, während es auf die einzelnen Massenteilchen nicht ankommt.

So führt man in der Hydrodynamik den Strömungsvektor ein, indem man den Mittelwert aus den Geschwindigkeiten der Moleküle bildet. Wie dieser Strömungsvektor, so haben alle die Ortsfunktionen (Skalare, Vektoren und Tensoren) in der mechanistischen Physik nur die Bedeutung mathematischer Hilfsmittel, welche zur Vereinfachung der Rechnungen dienen. Und ebenso kommt auch der partiellen Differentialgleichung hier keine selbständige physikalische, sondern nur eine rechnerisch-mathematische Bedeutung zu. Durch die Einführung dieser verschiedenen mathematischen Begriffsbildungen wird der Methode der mechanistischen Erklärung formale Eleganz und Übersichtlichkeit verliehen.

Nun hat sich aber doch der Standpunkt der Mechanistik nicht durchführen lassen; man ist genötigt worden, die Physik auf neue Grundlagen zu stellen und in den Begriffsbildungen noch viel weiter von der Wahrnehmung abzugehen, als es bei der Mechanistik geschieht. In der mechanistischen Theorie bleibt man trotz der weitgehenden Anwendung mathematischer Hilfsmittel und der Umdeutung aller Qualitäten doch der primitiven Auffassung insofern nahe, als die Grundannahmen, welche die Atome betreffen, ungefähr den Tatsachen entsprechen, die wir an großen Körpern wahrnehmen. In diesem Sinne hat also die mechanistische Theorie einen anschaulichen Charakter. Aber auch dieses Maß von Anschaulichkeit kann die Physik nicht aufrecht erhalten. Das Ideal, alle physikalischen Vorgänge durch die Mechanik zu umfassen, wird nicht erreicht. Es sind vor allem die Gebiete, welche die Strahlungsvorgänge betreffen, die Optik und die Theorie der Elektrizität, bei denen die Einordnung in die Mechanik nicht gelingt.

In der Optik hatte zwar die Wellentheorie des Lichtes große Erfolge, aber die Versuche, sich einen stofflichen Träger der Lichtschwingungen, den *"Äther"* in mechanischer Weise vorzustellen, führten zu immer künstlicheren Hypothesen. Die zunächst sich bietende Auffassung war die, den Äther als ein elastisches Medium zu betrachten, und zwar mußte er als ein starres, nur in dem kleinsten Teilen bewegliches Gerüst angenommen werden, bei welchem durch innere Verzerrungen elastische Kräfte hervorgerufen werden. Aus dieser Vorstellungsweise gewinnt man gemäß der Elastizitätstheorie ein System von partiellen Differentialgleichungen für die Komponenten u, v, w des Verrückungsvektors, als Funktionen der Ortskoordinaten x, y, z. Diese Differentialgleichungen ergeben sich formal durch Variation des Raumintegrals über die Energiedichte ε, welche sich als Funktion der partiellen Ableitungen von u, v, w nach x, y, z

$$u_x, u_y, u_z, v_x, v_z, w_x, w_y, w_z$$

in der Form

$$\begin{aligned}
&aH + bK;\\
&H = (u_x + v_y + w_z)^2\\
&K = (v_y - w_z)^2 + (w_z - u_x)^2 + (u_x - v_y)^2\\
&+ \frac{3}{2}\{(w_y + v_z)^2 + (u_z + w_x)^2 + (v_x + u_y)^2\}
\end{aligned}$$

darstellt (wobei a und b Konstanten bedeuten), und sie liefern in der Tat für ein homogenes Medium die optischen Gesetze. Jedoch bei der Untersuchung des Verhaltens der Wellen an der Grenze zweier Medien führt das Variationsprinzip nicht zu den tatsächlich bestehenden Beziehungen.

Man kann sich nun rein mathematisch die Frage vorlegen, wie das Variationsprinzip zu modifizieren ist, damit bei der Ausführung der Variation sich dieselben Differentialgleichungen, aber andere Randbedingungen ergeben, so daß also für das Innere eines Mediums die Bedingungen die gleichen bleiben, an der Grenze dagegen verändert werden. Die Variationsrechnung lehrt, daß man hierzu nur nötig hat, unter dem Integral einen Ausdruck von der Form

$$\frac{\partial A}{\partial x} + \frac{\partial B}{\partial y} + \frac{\partial C}{\partial z}$$

(die "Divergenz" eines Vektors mit den Komponenten A, B, C) hinzuzufügen. Es besteht nun die mathematische Identität:

$$4H - 2K + 3M = 12P,$$

worin H und K durch die obigen Ausdrücke gegeben sind und

$$M = (w_y + v_z)^2 + (u_z + w_x)^2 + (v_x + u_y)^2,$$

$$P = (v_y w_z - w_y v_z) + (w_z u_x - u_z w_x) + (u_x v_y - v_x u_y)$$

zu setzen ist; ferner ist

$$\begin{aligned} 2P &= \mathrm{div}(\mathrm{A}, \mathrm{B}, \mathrm{C}); \\ A &= u(v_y + w_z) - v u_y - w u_z \\ B &= v(w_z + u_x) - w v_z - u v_y. \end{aligned}$$

Dies besagt, daß der Ausdruck

$$-\frac{4}{3}H + \frac{2}{3}K$$

von M nur um eine Divergenz verschieden ist. Bei der Einführung von M als Integranden des Variationsprinzips bleiben also die Differentialgleichungen der Licht-Wellentheorie erhalten. Es zeigt sich ferner, daß diese Modifikation gerade die richtigen Randbedingungen liefert.

Indem man nun den Integranden des neuen Variationsprinzips wieder als Energiedichte ε ansieht, kommt man dazu,

$$\varepsilon = M = |\mathrm{rot}(\mathrm{u}, \mathrm{v}, \mathrm{w})|^2$$

anzunehmen. Dies ist der Ansatz der Lichttheorie von MacCullagh, nach welcher der Äther ein Stoff ist, der nicht der Verzerrung der kleinen Teile, dagegen ihrer Drehung einen Widerstand entgegensetzt. Eine solche Annahme ist aber sehr künstlich; man müßte, um den Widerstand gegen die Drehung mechanisch begreiflich zu machen, eigentlich zwei Äther annehmen, die sich durchdringen.

Auch die, insbesondere von Maxwell ausgeführten Versuche, durch eine Theorie der verborgenen Bewegungen die optischen und elektrodynamischen Gesetze auf mechanische Prozesse zurückzuführen, erwiesen sich als unfruchtbar. Noch gezwungener wurde die Äthertheorie, als Michelson seinen berühmten Interferenzversuch machte. Um diesen zu erklären, sah sich Lorentz zu der Annahme gedrängt, daß ein Körper, wenn er sich im Äther bewegt, in Richtung seiner Bewegung zusammenschrumpft.

Schließlich hat man den Gedanken, die Lichtschwingungen als einen mechanischen Vorgang zu erklären, ganz fallen gelassen und damit überhaupt das Ziel der Mechanistik preisgegeben. Man hat es als ein Vorurteil erkannt, daß die Grundgesetze der Physik den Gesetzen für große Körper analog sein müssen, und man ist hierin so weit gegangen, daß man nunmehr als das Elementare der physikalischen Vorgänge etwas annimmt, das gar nichts mehr zu tun hat mit dem, was wir direkt wahrnehmen.

Diese Entwicklung hat ihren Ausgang genommen von der durch Faraday begonnenen und durch Maxwell ausgebildeten *Feldtheorie* der Elektrodynamik. Hier werden die Fernkräfte gänzlich beseitigt, und auf den leeren (d. h. von Materie freien) Raum, dem vorher nur eine geometrische Bedeutung zukam, wird vor allem das Augenmerk gerichtet. Dieser wird nicht als absolut leer, sondern vielmehr als Sitz eines kontinuierlich ausgebreiteten "Feldes" angenommen, das sich mathematisch durch (örtlich und zeitlich veränderliche) Vektoren darstellt, den "elektrischen" Vektor $\mathcal{E}$ und den "magnetischen" $\mathcal{H}$. Diese sind miteinander verknüpft durch ein System von partiellen Differentialgleichungen; es sind das die Maxwellschen Gleichungen:

$$c \cdot \operatorname{rot} \mathcal{E} = -\frac{\partial \mathcal{H}}{\partial t}, \qquad \operatorname{div} \mathcal{H} = 0$$

$$c \cdot \operatorname{rot} \mathcal{H} = \frac{\partial \mathcal{E}}{\partial t}, \qquad \operatorname{div} \mathcal{E} = 0$$

worin c die Lichtgeschwindigkeit bedeutet. (In den beiden links stehenden Vektorformeln sind je drei Gleichungen zusammengefaßt.)

Charakteristisch für die Feldtheorie ist, daß hier die partiellen Differentialgleichungen nicht wie bei den mechanischen Theorien nur die Bedeutung von etwas mathematisch Abgeleitetem haben, sondern vielmehr die eigentliche Grundlage der Theorie bilden. Die Gültigkeit der Feldgleichungen bezieht sich auf den leeren Raum. An dessen Grenzen, wo die

Materie liegt, sind sie nicht mehr erfüllt. Somit werden die Orte der Materie als Stellen der Singularität des Feldes gekennzeichnet. Während man also vorher das Verständnis der physikalischen Erscheinungen durch Verfolgung der materiellen Teile anstrebte, sucht man es jetzt durch die Verfolgung der Feldzustände des leeren Raumes zu gewinnen. Das Feld wird als etwas physikalisch Wirkliches, aber nicht als der Materie gleichartig betrachtet. Die wesentliche Verschiedenheit von Feld und Materie zeigt sich mathematisch darin, daß die Grundgesetze für die Bestandteile der Materie die Form von *gewöhnlichen Differentialgleichungen* haben, wogegen die Grundgesetze des Feldes in *partiellen Differentialgleichungen* bestehen.

Es findet nun aber noch ein anderer wesentlicher Unterschied zwischen den Maxwellschen Feldgleichungen und den Grundgesetzen der Mechanik statt. Dieser betrifft die Frage der *Invarianz.* In Hinsicht auf diese ist zunächst zu bemerken, daß die mechanischen Gesetze gemeinsam mit den elektrodynamischen denjenigen Invarianzcharakter besitzen, welcher auf der zugrunde gelegten Euklidischen Geometrie beruht; das heißt sie bleiben ungeändert bei den kongruenten Raumtransformationen, welche durch Verschiebung und Drehung erzeugt werden, also den linearen Transformationen

$$\begin{aligned} x' &= a_{11}x + a_{12}y + a_{13}z + a_1 \\ y' &= a_{12}x + a_{22}y + a_{23}z + a_2 \\ z' &= a_{31}x + a_{32}y + a_{33}z + a_3 \end{aligned} \tag{1}$$

für welche identisch die Gleichung

$$\begin{aligned} &(a_{11}x + a_{12}y + a_{13}z)^2 + (a_{21}x + a_{22}y + a_{23}z)^2 \\ &+ (a_{31}x + a_{32}y + a_{33}z)^2 = x^2 + y^2 + z^2 \end{aligned}$$

besteht (und die Determinante

$$\begin{vmatrix} a_{11} & a_{12} & a_{13} \\ a_{21} & a_{22} & a_{23} \\ a_{31} & a_{32} & a_{33} \end{vmatrix}$$

positiv ist). Um diese Invarianz der Gesetze deutlich hervortreten zu lassen, wendet man die Vektorenrechnung an, mit Hilfe deren man es einem Gleichungssystem sofort ansehen kann, wenn es gegenüber den Transformationen (1) invariant ist.

Bei den Grundgesetzen der Mechanik kommt zu dieser räumlichen Invarianz noch diejenige gegenüber einer gleichförmigen Translation des Bezugssystems; die Grundgleichungen der Mechanik bleiben formal ungeändert bei einer Transformation der Form

$$\begin{aligned} x' &= x + v_1 t \\ y' &= y + v_2 t \\ z' &= z + v_3 t \end{aligned} \tag{2}$$

denn bei einer solchen bleiben die räumlichen Entfernungen sowie die Beschleunigungen ungeändert, und anderweitig treten die Raum- und Zeitgrößen in den mechanischen Bewegungsgleichungen nicht auf. Diese Invarianz ist der mathematische Ausdruck des *Galileischen Trägheitsprinzips*, gemäß welchem eine geradlinig gleichförmige Bewegung eines Systems durch mechanische Vorgänge innerhalb des Systems nicht feststellbar ist.

Die Vereinigung der Transformationen (1) und (2) ergibt eine Transformation

$$\begin{aligned} x' &= a_{11}x + a_{12}y + a_{13}z + v_1 t + a_1 \\ y' &= a_{21}x + a_{22}y + a_{23}z + v_2 t + a_2 \\ z' &= a_{31}x + a_{32}y + a_{33}z + v_3 t + a_3 \end{aligned} \tag{3}$$

wobei $a_{11}, \ldots, a_{33}$ den vorhin genannten Bedingungen genügen. Ebenso wie die Grundgesetze der Mechanik müssen auch alle aus der mechanischen Theorie abgeleiteten Gesetze bei diesen Transformationen ungeändert bleiben.

Die Maxwellschen Gleichungen besitzen hingegen nicht diese Invarianz, vielmehr ist es eine andere Art von linearen Raum-Zeit-Transformationen, welche die Maxwellschen Gleichungen ungeändert lassen, nämlich die *“Lorentz-Transformationen”*:

$$\begin{aligned} x' &= a_{11}x + a_{12}y + a_{13}z + v_1 t + a_1 \\ y' &= a_{21}x + a_{22}y + a_{23}z + v_2 t + a_2 \\ z' &= a_{31}x + a_{32}y + a_{33}z + v_3 t + a_3 \\ t' &= a_{41}x + a_{42}y + a_{43}z + a_{44} t + a_4 \end{aligned} \tag{4}$$

bei denen identisch die Gleichung

$$\begin{aligned} &(a_{11}x + a_{12}y + a_{13}z + a_{14}t)^2 + (a_{21}x + a_{22}y + a_{23}z + a_{24}t)^2 \\ &+(a_{31}x + a_{32}y + a_{33}z + a_{34}t)^2 - c^2(a_{41}x + a_{42}y + a_{43}z + a_{44}t)^2 \\ &\quad = x^2 + y^2 + z^2 - c^2t^2 \end{aligned}$$

erfüllt ist (und die Determinante

$$\begin{vmatrix} a_{11} & a_{12} & a_{13} & a_{14} \\ a_{21} & a_{22} & a_{23} & a_{24} \\ a_{31} & a_{32} & a_{33} & a_{34} \\ a_{41} & a_{42} & a_{43} & a_{44} \end{vmatrix}$$

positiv ist). Der charakteristische Unterschied dieser Transformationen gegenüber den Transformationen (3) besteht darin, daß hier die Zeit t mit transformiert wird.

Es standen nun zunächst die Mechanik und die Maxwellsche Elektrodynamik als zwei getrennte Gebiete der Physik nebeneinander, jedes mit einem anderen Invarianzprinzip verbunden. Dieser systematisch unbefriedigende Zustand konnte sich nicht lange halten. Durch experimentelle Ergebnisse im Zusammenhang mit theoretischen Erwägungen wurde man dazu gedrängt, dem Invarianzprinzip der Feldtheorie universale Gültigkeit zuzuschreiben. Von Einstein wurde das Postulat aufgestellt, daß alle Naturgesetze gegenüber den Lorentz-Transformationen invariant sein müssen. Diesem neuen Prinzip mußte die alte Mechanik weichen.

Zu dieser Umwälzung trug wesentlich die Entwicklung der Elektrizitätstheorie bei. Hier bildete man sich anfangs zur mechanistischen Erklärung die Vorstellung eines elektrischen Fluidums. Die neueren experimentellen Erfahrungen, insbesondere die Untersuchungen über die Entladungen in evakuierten Röhren führten dann zu der Auffassung, daß die Elektrizität atomistisch beschaffen ist. Die Annahme von *Elektronen* (elektrisch geladenen Teilchen) als letzten Bestandteilen der Materie erwies sich als äußerst fruchtbar, und die mechanistischen Vorstellungen führten in der Elektronentheorie zu bedeutenden Erfolgen.

Es zeigte sich jedoch, daß die gewöhnlichen Gesetze der Mechanik auf die Bewegung der Elektronen nicht anwendbar sind, indem eine *Abhängigkeit der Trägheit von der Geschwindigkeit* gefunden wurde. Für diese Tatsache erhält man nun eine überraschend einfache Erklärung, wenn man das Grundgesetz der Mechanik

$$\text{Kraft} = \text{Masse} \times \text{Beschleunigung}$$

im Sinne des Invarianzprinzips der Elektrodynamik modifiziert. Betrachten wir den einfachsten Fall, eine Bewegung in der x-Achse unter der Wirkung einer konstanten Kraft, so lautet das Bewegungsgesetz der Mechanik:

$$m \cdot \frac{d^2x}{dt^2} = K.$$

Diese Gleichung ist invariant gegenüber der Transformation

$$x' = x + vt,$$

da ja

$$\frac{dx'}{dt} = \frac{dx}{dt} + v$$

$$\frac{d^2x'}{dt^2} = \frac{d^2x}{dt^2}$$

ist. Sie bleibt dagegen nicht ungeändert bei den Lorentz-Transformationen, die in dem betrachteten Spezialfall die Form

$$x' = \frac{x + vt}{\sqrt{1 - \frac{v^2}{c^2}}} \qquad t' = \frac{\frac{v}{c^2}x + t}{\sqrt{1 - \frac{v^2}{c^2}}} \tag{5}$$

haben. Wie man sieht, geht die Transformation (5) für $c = \infty$ in die alte Transformation

$$x' = x + vt$$
$$t' = t$$

über. Soll nun die Bewegungsgleichung so abgeändert werden, daß sie gegenüber Lorentz-Transformationen invariant ist, so muß anstelle von d^2x/dt^2 eine Invariante der Transformation (5) gesetzt werden: und zwar wird man eine solche wählen, die für $c = \infty$ in d^2x/dt^2 übergeht (so daß d^2x/dt^2 als ein angenäherter Ausdruck betrachtet werden kann). Eine derartige Invariante ist

$$\frac{d^2x}{dt^2} \cdot \frac{1}{\left(1 - \frac{1}{c^2}\left(\frac{dx}{dt}\right)^2\right)^{\frac{3}{2}}}$$

und man kommt also zu der Annahme, daß die richtige Formel des Bewegungsgesetzes

$$\frac{m}{\left(1 - \frac{1}{c^2}\left(\frac{dx}{dt}\right)^2\right)^{\frac{3}{2}}} \cdot \frac{d^2x}{dt^2} = K$$

lautet. Diese Gleichung erhält wieder die Form des alten Gesetzes, wenn man als träge Masse den Ausdruck

$$\frac{m}{\left(1 - \frac{1}{c^2}\left(\frac{dx}{dt}\right)^2\right)^{\frac{3}{2}}}$$

nimmt. Hierdurch gelangt man zu der Folgerung, daß die träge Masse von der Geschwindigkeit abhängt, und es stellt sich nun heraus, daß diese Abhängigkeit gerade so ist, wie sie sich auf experimentellem Wege ergeben hat.

So wie hier bei der Elektronenmechanik führt noch auf verschiedenen Gebieten das elektrodynamische Invarianzprinzip zu verblüffend einfachen Erklärungen von vorher unverstandenen experimentellen Ergebnissen (Fizeauscher Versuch, Elektrodynamik bewegter Körper), wobei deutlich hervortritt, daß für diese Erklärungen gerade die Abweichung der Lorentz-Transformation (4) von der gewöhnlichen Galileischen Transformation (3) das Entscheidende ist.

Aufgrund dieser Erkenntnisse hat sich die Überzeugung durchgesetzt, daß die Art der Invarianz, welche den Maxwellschen Gleichungen zukommt, überhaupt für alle Naturgesetze besteht. Und so ist die Feldtheorie zu voller

Herrschaft gelangt, während die Mechanik ihre grundsätzliche Bedeutung verloren hat, indem sie nur noch als eine Annäherung betrachtet wird, deren Anwendbarkeit darauf beruht, daß die körperlichen Geschwindigkeiten im Vergleich zur Lichtgeschwindigkeit sehr gering sind.

Durch die Annahme des neuen Invarianzprinzips wird das physikalische Denken, ja sogar auch die Auffassung von Raum und Zeit von Grund aus umgestaltet. Es kann jetzt nur das als eine objektive Bestimmung der Natur anerkannt werden, was gegenüber Lorentz-Transformationen einen invarianten Charakter hat. Und wie mit Hilfe der bisherigen Vektoranalysis die physikalische Begriffsbildung dem Invarianzgesetz der kongruenten Raumtransformationen (1) angepaßt wurde, so muß nun durch Ausbildung einer auf das Vierdimensionale bezüglichen Vektoranalysis die Invarianz gegenüber Lorentz-Transformationen zur Geltung gebracht werden.

Bei dieser neuen Darstellung der physikalischen Gesetze ergeben sich ganz erhebliche, überraschende Vereinfachungen. Es fällt uns wie Schuppen von den Augen, wenn wir sehen, wie sich da alles fügt. Das modifizierte Trägheitsgesetz wird mit dem Energiesatz der Mechanik zu einer Vierervektorgleichung vereinigt; der Satz von der Erhaltung der Masse (Kontinuitätsgleichung), ebenso wie der von der Erhaltung der elektrischen Ladung nimmt die einfache Form an, daß eine Viererdivergenz gleich Null ist. Der Zusammenhang zwischen Elektrizität und Magnetismus erhält eine mathematische Deutung, indem die elektrische und die magnetische Feldstärke zu einem Sechservektor vereinigt wird. (Sechs ist die Zahl der Projektionen eines Flächenstückes auf die zweidimensionalen Koordinaten-Ebenen). Hierbei bekommt das System der Maxwellschen Gleichungen eine erstaunlich einfache Gestalt: Aus den partiellen Ableitungen des elektromagnetischen Sechservektors werden nach den einfachsten Bildungsgesetzen der Vierervektoranalysis zwei Vierervektoren zusammengesetzt; das Verschwinden dieser Vektoren (d. h. ihrer Komponenten) ist der Inhalt der Maxwellschen Gleichungen, im Falle des Vakuums. (Werden elektrische Ladungen mit in Betracht gezogen, so stellt sich die Aussage der Maxwellschen Theorie so dar, daß der eine der beiden Vierervektoren gleich der "Viererdichte" der elektrischen Ladung, der andere gleich Null zu setzen ist). Die elektromagnetische Energiedichte wird mit dem Pointingschen Strahlungsvektor und den Maxwellschen Spannungen zu einem einzigen invarianten Gebilde mit $4 \cdot 4$ Komponenten, einem Sechzehnertensor zusammengefaßt. (Zufolge seiner Symmetrie wird dieser Tensor bereits durch zehn Komponenten charakterisiert.)

Mit diesen Begriffsbildungen entfernt sich die Physik gänzlich von dem Anschaulichen, nachdem bereits der Begriff des Feldes ein Abgehen von der geläufigen Vorstellung des physikalischen Stoffes (des Körperlichen) bedeutete. Der Gedanke der Feldtheorie in Verbindung mit dem neuen Invarianzprinzip kennzeichnet eine *zweite Stufe* in der Entwicklung der Physik, deren Voraussetzungen und Ziele nunmehr völlig andere geworden sind als auf dem Stadium der Mechanik.

Im Sinne der Leitgedanken dieser neuen Richtung der Physik mußte nun noch ein bedeutsamer Schritt getan werden, der wiederum eine Umwälzung in den Grundlagen der Physik mit sich gebracht hat. Es weist nämlich die Gestaltung der Physik, soweit wir sie bisher verfolgt haben, in Hinsicht auf das Ziel einer einheitlichen Feldtheorie noch zwei wesentliche Unvollkommenheiten auf. Erstens steht das Gravitationsgesetz als ein Überbleibsel aus der Mechanik ganz abseits von der übrigen Physik. Zweitens besteht ein Kontrast insofern, als das Invarianzgesetz in seinem mathematischen Charakter von ganz anderer Art ist als die Feldgesetze. Während nämlich die Feldgleichungen Differentialgesetze sind, welche Beziehungen im Infinitesimalen ausdrücken, stellen die Formeln der Lorentz-Transformation ein Ferngesetz dar, welches auf der ferngesetzlichen Euklidischen Geometrie beruht. Hier wird die geradlinig gleichförmige Bewegung (die Gerade der vierdimensionalen Welt) von vornherein ausgezeichnet, und damit hängt zusammen, daß die physikalischen Gleichungen sich nur auf gewisse spezielle Koordinatensysteme (Inertialsysteme) beziehen. Auch wirkt es befremdlich, daß für die Beziehungen von Raum und Zeit die Konstante c so eine ausgezeichnete Rolle spielt.

Diese Mängel verschwinden in der von Einstein in neuester Zeit aufgestellten Gravitationstheorie. Einstein geht hier von der Forderung aus, daß alle Naturgesetze *gegenüber beliebigen Transformationen invariant* sein müssen; beim Übergang von einem Koordinatensystem (x, y, z, t) zu einem anderen durch eine beliebige Transformation

$$
\begin{aligned}
x' &= X(x, y, z, t)\\
y' &= Y(x, y, z, t)\\
z' &= Z(x, y, z, t)\\
t' &= T(x, y, z, t)
\end{aligned}
$$

(die nur gewissen Bedingungen der Stetigkeit, Differenzierbarkeit und Umkehrbarkeit zu genügen hat), sollen die Naturgesetze in ihrer mathematischen Form ungeändert bleiben.

Der Sinn dieses allgemeinen "Relativitätsprinzips" ist, daß eine von aller Subjektivität und Willkür befreite Darstellung der Naturzusammenhänge nur eine solche sein kann, die unabhängig ist von der Art, wie die Weltpunkte (durch Koordinaten) bezeichnet werden. Damit eine solche invariante Darstellung der Naturgesetze möglich ist, müssen die Methoden der Differentialgeometrie angewandt und die geometrischen Beziehungen mit den physikalischen zu einer Einheit verschmolzen werden. Wie in der Differentialgeometrie gelehrt wird, hängt die Maßbestimmung (Längenmessung,

Winkelmessung usw.) einer allgemeinen vierdimensionalen Mannigfaltigkeit bei Zugrundelegung beliebiger Koordinaten von einem Funktionensystem

$$g_{\mu\nu}(x,y,z,t); \quad \mu,\nu = 1,2,3,4$$
$$(g_{\mu\nu} = g_{\nu\mu})$$

ab, welche beim Übergang zu anderen Koordinaten einem bestimmten Transformationsgesetze unterliegt.

Bei einer invarianten Formulierung der Naturgesetze muß nun das Funktionensystem $g_{\mu\nu}$, durch welches sich die Metrik der vierdimensionalen Raum-Zeit-Mannigfaltigkeit bestimmt, in den physikalischen Gleichungen auftreten. Diesen Sachverhalt deutet Einstein so, daß er das System der $g_{\mu\nu}$ als den Ausdruck von etwas Physikalischem, des "metrischen Feldes" auffaßt. Dieses Feld bestimmt außer der Längen- und Winkelmessung auch die Art der Lichtausbreitung (nach Weg und Geschwindigkeit); und aus den partiellen Ableitungen der $g_{\mu\nu}$ setzen sich die Kräfte zusammen, welche die Erscheinungen der *Trägheit* und der *Schwere* ausmachen. Den Funktionen $g_{\mu\nu}$ kommt also insbesondere die Bedeutung von *Gravitationspotentialen* zu.

Durch diesen Ansatz der Einsteinschen Theorie wird somit das Gesetz der Gravitation zu einem Feldgesetz, welches mit der Maßbestimmung der vierdimensionalen Welt in unlösbarer Verknüpfung steht. Die Einheit der Theorie verlangt nun noch einen Zusammenhang zwischen dem metrischen und dem elektrodynamischen Feld. Dieser wird durch einen weiteren prinzipiellen Gedanken hergestellt, mit welchem Einstein an Riemann anknüpft. Während bisher die Maßbestimmung der Raum-Zeit-Mannigfaltigkeit von vornherein als so beschaffen vorausgesetzt wurde, daß bei geeigneter Koordinatenwahl die $g_{\mu\nu}$ von Ort und Zeit unabhängig sind, nimmt Einstein an, daß das metrische Feld durch die "Materie" bestimmt wird, wobei er sich die Materie als elektromagnetisch konstruiert denkt, so daß ihre räumlich-zeitliche Verteilung durch die elektromagnetischen Feldgrößen charakterisiert wird. Diese Annahme besagt, daß es ein System von Feldgleichungen gibt, gemäß welchen die $g_{\mu\nu}$ von dem elektromagnetischen Felde abhängen. Ein solches Gleichungssystem hat Einstein wirklich aufgestellt und er konnte zeigen, daß dieses in erster Annäherung zu dem Newtonschen Gravitationsgesetz und bei weiterer Annäherung zu einer der Erfahrung entsprechenden Modifikation des Newtonschen Gesetzes führt.

In dieser Einsteinschen Theorie haben wir nun eine konsequente Feldtheorie vor uns; die zweite Stufe in der Entwicklung der Physik ist damit erreicht. Nicht bloß eine Ausschaltung der Sinne, wie bei der Mechanistik, findet hier statt, sondern eine gänzliche Beseitigung des Anthropomorphismus. Die Begriffsbildungen haben sich ganz und gar von dem anschaulich Geläufigen emanzipiert; und gerade dadurch, daß man sich von der Anschauung los macht, wird die Objektivität in der Auffassung der Naturgesetze sowie die Einheit und Übersichtlichkeit des theoretischen Systems erreicht. In dieser Hinsicht möchte ich das allgemeine Relativitätsprinzip als den höchsten Triumph des Geistes über die Erscheinungswelt ansehen.

5. KAPITEL: DIE GESETZE DER PHYSIK UND EWIGE NATURGESETZE

Ich gehe nun zu dem zweiten Gesichtspunkt über, hinsichtlich dessen die primitive Auffassung von der Physik zu modifizieren ist. Gemäß dem logischen Schema "Begriff, Urteil, Schluß" handelt es sich dabei um die physikalischen Urteile, also um die *Gesetze*, welche die Physik aufstellt. Nach der landläufigen Ansicht besteht die physikalische Forschung im Auffinden der unabänderlichen Naturgesetze. Danach sollte man denken, daß die in der Physik gefundenen Gesetze, so wie die mathematischen Sätze, einen dauernden Bestand feststehender Wahrheiten bilden, der sich beim Fortschreiten der physikalischen Erkenntnis immer nur erweitert.

Dies trifft aber durchaus nicht zu. Allerdings müssen wir die Naturgesetze als ewige Gesetze ansehen. In dem Gedanken der Naturgesetzlichkeit ist bereits die Unabänderlichkeit der Gesetze enthalten. Würden heute diese, morgen jene Gesetze gelten, so hieße das, daß überhaupt keine Gesetzlichkeit im strengen Sinne besteht. Die Behauptung einer Gesetzlichkeit, die sich nur auf einen bestimmten Zeitraum bezieht, würde überhaupt kein Gegenstand exakter wissenschaftlicher Kontrolle sein. Denn in einem beschränkten Zeitintervall können wir nur endlich viele Beobachtungen anstellen, und nach Ablauf der betreffenden Zeit könnte zur experimentellen Untersuchung des Gesetzes nichts mehr beigetragen werden. Wir hätten also zur Prüfung des Gesetzes nur eine abgegrenzte endliche Gesamtheit von Daten zur Verfügung. Durch eine solche kann aber die Gültigkeit eines Gesetzes nicht entschieden werden. Denn durch einen endlichen Tatsachenkomplex wird ein Gesetz—das ja seinem Sinne nach unendlich viele Fälle umfaßt—niemals eindeutig festgelegt. Ein und dieselbe Gesamtheit von Beobachtungen kann auf verschiedene Weisen durch Interpolation in ein Gesetz gefaßt werden, und sofern nicht weitere Erfahrungen hinzukommen, hat es keinen Sinn, eines von diesen Gesetzen als das wahrhaft gültige zu erklären. Den Naturgesetzen müssen wir also unabänderliche Gültigkeit zuschreiben, sofern wir überhaupt an der Grundannahme des Physikers festhalten, daß es Naturgesetze gibt.

Nun sind aber die in der Physik aufgestellten Gesetze keineswegs identisch mit den ewigen Naturgesetzen, und sie besitzen nicht wie diese den Charakter der Unabänderlichkeit. Vielmehr finden wir, daß die zunächst aufgestellten Gesetze nachher korrigiert werden und sich dann als Folge einer allgemeineren Gesetzlichkeit herausstellen. Der Grund hierfür besteht in folgendem. Das Suchen von Gesetzen ist vergleichbar mit einem Erraten. Dabei liegt nun der komplizierende Umstand vor, daß die Anhaltspunkte, welche die Beobachtungen für dieses Erraten liefern, nicht alle auf einmal gegeben sind, sondern sich erst im Laufe der Entwicklung der Wissenschaft vervollständigen. Die Gesetze werden daher nicht mit einem Schlage, sondern nur auf dem Wege einer schrittweisen Annäherung gefunden.

Es ist also die Unvollständigkeit und Ungenauigkeit unserer Beobachtungen, welche sich hier geltend macht. Diese bildet einerseits einen Übelstand, andererseits kommt sie uns doch auch bei der Auffindung der Gesetze zustatten. Wir würden bei vollkommen genauer Beobachtung durch die Mannigfaltigkeit und Komplikation der Erscheinungen gänzlich erdrückt und kämen gar nicht zur Aufstellung von Gesetzen. So wäre z. B.—wie Poincaré einmal bemerkte—Kepler nicht zu seinen Gesetzen gekommen, wenn er ganz genau beobachtet hätte. Der Fortschritt der physikalischen Erkenntnis ist also mit einer beständigen Modifikation der Gesetze verknüpft, die zur Erklärung der jeweils bekannten Tatsachen angenommen werden. Ich will dies an einigen Beispielen darlegen.

Die *Keplerschen Gesetze* der Planetenbewegung hielt man zunächst für ganz streng und bewunderte gerade in der Einfachheit der Planetenbahnen die Harmonie der Natur. Seitdem hat man erkannt, daß die Kegelschnittbahnen nur eine Näherung bilden. Diese Auffassung ergab sich mit Notwendigkeit, sobald man gefunden hatte, daß die Keplerschen Gesetze aus dem allgemeinen Newtonschen Gravitationsgesetz als Folgerung abzuleiten sind.

Das *Boyle-Mariottesche Gesetz*,

$$p \cdot v = \mathrm{const} \cdot \theta ,$$

worin p den Druck, v das spezifische Volumen und θ die Temperatur eines Gases bedeutet, wurde zunächst schlechthin als Zustandsgleichung für die Gase aufgestellt. Danach fand man, daß es nur angenähert für hohe (d. h. hinreichend weit über der kritischen Temperatur des Gases liegende) Temperaturen gelte und daß es durch die (auch auf Flüssigkeiten anwendbare) van der Waalssche Gleichung zu korrigieren sei. Dieses Gesetz hat sich dann wiederum als nicht vollkommen genau erwiesen. Ferner erkannte man, daß die thermischen Zustandsgleichungen sich aus der kinematischen Gastheorie ableiten lassen, wobei sich die aufgestellten Formeln als Näherungen ergeben.

In der Theorie der Elektrizität glaubt man eine Zeitlang in dem *Weberschen Gesetz*, durch welches zum ersten Mal statische und galvanische Elektrizität theoretisch zusammengefaßt wurden, das strenge Elementargesetz der elektrischen Wirkungen gefunden zu haben. Durch die Maxwellsche Theorie wurde dieses Gesetz vollständig überholt und stellt sich nunmehr als eine nur grobe Annäherung dar, die gar keinen Einblick in die wesentlichen Zusammenhänge bietet.

Nun könnte man sagen, daß bei allen den genannten Fällen die Gesetze teils einen zu speziellen Charakter haben, teils ihrer mathematischen Form nach nicht einfach genug sind, um als endgültiger Ausdruck von Naturgesetzen im strengen Sinne gelten zu können, und dies sei der Grund, weshalb jene Gesetze so kurzlebig sind. Demgegenüber will ich an das Gravitationsgesetz erinnern, das universellen Charakter besitzt und durch die Einfachheit seiner Form ausgezeichnet ist, das ferner in allen seinen Konsequenzen sich mit astronomischer Genauigkeit bestätigt hat und durch Generationen hindurch für etwas Unabänderliches gehalten wurde. Dieses Gesetz erweist sich in der Einsteinschen Theorie als nicht genau gültig. Es wird verdrängt durch ein Gesetz von höherer Allgemeinheit, welches mit dem früheren nicht die entfernteste Ähnlichkeit hat. Nur aufgrund der besonderen Beschaffenheit des uns umgebenden Feldes, d. h. zufolge des Umstandes, daß die Materie im Weltraum so dünn verteilt ist, kommt das Newtonsche Gesetz als erste Annäherung heraus. In zweiter Annäherung ist das Newtonsche Gesetz nicht mehr richtig, vielmehr ergibt sich aus dem Einsteinschen Gesetz eine Korrektur, welche bei der Perihelbewegung des Merkur zur Geltung kommt. Das Fortschreiten des Merkurperihels, welches bei der Einsteinschen Theorie direkt aus dem Gravitationsgesetz folgt, war bereits als Tatsache bekannt, es ließ sich aber aus der Newtonschen Theorie nicht ohne besondere zusätzliche Hypothesen erklären. Dieser Umstand allein hätte jedoch nicht dazu genötigt, das alte Gravitationsgesetz preiszugeben. Sondern es waren grundsätzliche Erwägungen, welche dazu geführt haben.

Bei der *Entwicklung der Chemie* verhält es sich dagegen so, daß durch rein experimentelle Entdeckungen die Grundlagen der Theorie umgestoßen wurden. Die Unveränderlichkeit, Unteilbarkeit und Unzerstörbarkeit der Atome wurde seit langen Zeiten als feststehende Tatsache angenommen, und diese Annahme bildete bisher die Grundlage für die Chemie. Nun ist durch die neuerdings entdeckten Erscheinungen der Radioaktivität diese Voraussetzung gänzlich unhaltbar geworden, und an ihrer Stelle ist eine ganz neue theoretische Auffassung getreten. Nach dieser sind die Atome aus kleineren Bestandteilen zusammengesetzt; je nachdem die Verbindung dieser Teile stabil oder labil ist, bleibt das Atom beständig oder es zerfällt. Beim Zerfall der Atome entstehen aus einem Stoff andere Stoffe, die in ihren Eigenschaften (Atomgewicht, Schmelzpunkt, Siedepunkt) voneinander grundverschieden sind. So verwandelt sich z. B. Radium, das bei gewöhlicher Temperatur im festen Zustande ist, in die gasförmige "Radium-Emanation"; und nach mehreren weiteren Umwandlungen entsteht ein dem Blei sehr verwandter Stoff. Alle zerfallenden ("radioaktiven") Elemente senden "α-Strahlen" aus, die aus Helium-Atomen bestehen. Diese Anschauungen haben in jüngster Zeit noch dadurch eine Bestätigung erfahren, daß es gelungen ist, künstlich die Zerstörung von Stickstoff-Atomen herbeizuführen. So verliert der Begriff des chemischen Elementes ganz und gar seine prinzipielle Bedeutung, und man kommt zu der Vermutung, daß alle Stoffe aus (positiv elektrischen) Helium-Atomen und (negativen) Elektronen aufgebaut sind.

Diese zwei Beispiele aus der neuesten Entwicklung der Physik zeigen deutlich, wie auch solche Gesetze der Physik, die man als etwas ganz Festes und Endgültiges betrachtet, umgestoßen werden können. Doch die Änderung in den Gesetzen der Physik geht noch viel tiefer. Sie betrifft sogar solche Gesetze, die man früher gar nicht zur Physik gerechnet hat, die man vielmehr als selbstverständliche Voraussetzungen jeder Physik annahm. Nämlich auch in den vorphysikalischen geometrisch-kinematischen Grundlagen der Physik hat sich eine starke Wandlung der Auffassung vollzogen. Der Satz vom Parallelogramm der Bewegungen (die Regel für die Addition der Geschwindigkeiten) wird heute nicht mehr anerkannt, die Begriffe der Gleichzeitigkeit und der körperlichen Gestalt haben ihre eigentliche Bedeutung verloren, indem sie nur relativ zu einem Koordinatensystem Anwendung finden, dessen Wahl ja von dem jeweiligen Zweck der Untersuchung abhängt. Also selbst solche Sätze, wie die Behauptung der absoluten (objektiven) Bedeutung der Gleichzeitigkeit sind keine ewigen Naturgesetze.

Hiernach könnte es scheinen, als ob in der Physik überhaupt alles dem Wandel unterworfen sei. Das ist aber doch nicht der Fall, vielmehr gibt es physikalische Prinzipien, denen eine dauernde Bedeutung zugeschrieben werden kann. Ein solches Prinzip ist der *Entropiesatz.* Dieser ist zwar in der Form, wie er zunächst aufgestellt wurde, nicht streng gültig, vielmehr hat sich gezeigt, daß er nur im Sinne eines Wahrscheinlichkeitssatzes zutrifft. In dieser Bedeutung wird er aber seine Rolle in all den Gebieten der Physik, wo die Methode der Statistik zur Geltung kommt, dauernd fortspielen, und es erscheint undenkbar, daß wir durch Theorie oder Erfahrung genötigt werden sollten, ihn fallen zu lassen.

Ferner ist hier der *Satz von der Erhaltung der Energie* zu nennen, welcher in der neueren Physik eine der fruchtbarsten Entdeckungen bildet. Bei der mechanistisch-atomistischen Auffassung ist er eine strenge Folge aus dem Grundgesetz, und er dient daher als eine Stütze für diese Auffassung. Aber auch auf der zweite Stufe behält der Energiesatz seine Gültigkeit, indem er sich als Folge der Feldgleichungen (in der Form des Verschwindens einer Tensordivergenz) ergibt. Ja er gewinnt hier sogar eine Erweiterung und Vertiefung, indem die Masse als ein besondere Form der Energie erscheint und somit der Satz von der Erhaltung der Masse in dem Satz von der Erhaltung der Energie aufgeht. So hat sich der Energiesatz als dauerndes Prinzip der Physik erhalten, und bei den Modifikationen seiner Auffassung hat sich der Bereich seiner Gültigkeit immer mehr erweitert.

Endlich das *allgemeine Relativitätsprinzip* ist nicht wieder aus der Welt zu schaffen. Wenn irgend von einem ewigen Naturgesetz die Rede sein kann, dann muß dieses als ein solches angesprochen werden.

6. KAPITEL: DIE BEZIEHUNG ZWISCHEN THEORIE UND EXPERIMENT

Wir kommen nun in der Berichtigung der primitiven Auffassung über die Physik zu dem dritten Punkt, welcher das Schließen betrifft. Da sich die Schlüsse der Physik auf Erfahrungstatsachen stützen, so haben wir hier das Verhältnis zwischen *Theorie und Experiment* zu betrachten. Dieses besteht nach der landläufigen Meinung darin, daß der Physiker zunächst Hypothesen bildet und daß zwischen diesen dann das Experiment entscheidet. In Wirklichkeit ist aber die Beziehung zwischen Hypothese und Experiment nicht so einfach und eindeutig. Insbesondere trifft es nicht zu, daß das Experiment immer erst das Nachträgliche ist. Viele Experimente werden auf Vorrat gemacht. So werden mannigfache experimentelle Bestimmungen von physikalischen Konstanten (z. B. Materialkonstanten) ausgeführt, es werden Tabellen angelegt, z. B. für die Spektren der verschiedenen Stoffe, ohne daß man dabei eine Entscheidung über bestimmte Hypothesen im Auge hat. Auch gibt es ja experimentelle Entdeckungen, welche gar nicht durch die zu ihrer Zeit vorliegenden physikalischen Hypothesen vorbereitet sind, vielmehr dem physikalischen Denken ganz neue Bahnen weisen und neue Gebiete erschließen. Es geht also in der Entwicklung der physikalischen Wissenschaft nicht immer die Hypothesenbildung voran, sondern teils hat die Theorie, teils aber auch das Experiment die führende Rolle. Dies wollen wir nun genauer an Beispielen verfolgen, und zwar haben wir vier typische Fälle zu unterscheiden.

Der erste Fall ist derjenige, welchen die anfangs formulierte landläufige Ansicht allein ins Auge faßt, wo die Theorie zu Folgerungen führt, die sich experimentell kontrollieren lassen und durch die Ausführung des Experimentes die Entscheidung für oder gegen die Theorie herbeigeführt wird. Ein klassisches Beispiel hierfür findet sich in der Geschichte der Optik. Hier standen sich anfangs zwei Theorien gegenüber: die *Emissionstheorie von Newton* und die *Wellentheorie von Huygens*, welche beide die Gesetze der Reflexion und der Brechung des Lichtes erklärten. Bei der Erklärung der Brechung macht sich nun der Unterschied der beiden Theorien dadurch geltend, daß nach der Newtonschen Auffassung die Fortpflanzungsgeschwindigkeit des Lichtes in dem optisch dichteren Medium (wo der Lichtstrahl nach der Normale der Grenzfläche hin gebrochen wird) größer sein muß, also z. B. im Wasser größer als in Luft, während es sich nach Huygens umgekehrt verhält. Auf diese Weise ergab sich die Möglichkeit einer experimentellen Entscheidung durch die Bestimmung der Lichtgeschwindigkeit im Wasser und in Luft. Diese hat Foucault ausgeführt, und sein Experiment sprach zugunsten der Huygensschen Theorie.

Berühmt ist ferner die Voraussetzung der Existenz des *Neptun* durch Adams und Leverrier. Diese erklärten die beobachteten Störungen in der

Bahn des Uranus gemäß der Newtonschen Gravitationstheorie durch die Annahme eines vordem unbekannten Planeten, der dann auch an der vorausbestimmten Stelle entdeckt wurde.

Eine Folgerung der kinetischen Gastheorie ist, daß für ein einatomiges Gas das Verhältnis der *spezifischen Wärmen*

$$\frac{c_p}{c_v} = \frac{5}{3} \qquad \text{ist.}$$

(c_p bedeutet die spezifische Wärme bei konstantem Druck, c_v bei konstantem Volumen). Diese Beziehung hat man bei Versuchen an Quecksilberdampf (welcher einatomig ist) bestätigt gefunden.

Die Existenz der *elektrischen Wellen*, auf denen die drahtlose Telegraphie beruht, wurde durch die Maxwellsche Theorie vorausgesagt. Es gelang dann Hertz, diese Wellen experimentell aufzuweisen, und damit erhielt die Maxwellsche Theorie eine glänzende Bestätigung.

Ein modernes Beispiel bildet das *Experiment von Laue*. Es lag die Hypothese vor, daß die Röntgenstrahlen in einer Wellenausbreitung von derselben Art wie die Lichtwellen bestehen; und zwar folgte aus der Theorie, daß die Länge dieser Wellen die Größenordnung 10^{-9} hat. Andererseits lehrte die Kristalltheorie, daß die Atome in den Kristallen gitterförmig gelagert sind, und man konnte die Abstände der Atome ermitteln. Es ergab sich dabei für alle Kristalle die Größenordnung der Abstände gleich 10^{-8}. Danach sind also die Abstände der Atome in einem Kristall nahezu von derselben Größenordnung wie die Wellenlänge der Röntgenstrahlen. Hieraus zog nun Laue den Schluß, daß ein Kristall sich als Beugungsgitter für Röntgenstrahlen verwenden lassen müße; er führte nun das Experiment der Durchstrahlung eines Kristalles mit Röntgenstrahlen wirklich aus, und es zeigten sich in der Tat die erwarteten Interferenzerscheinungen.

Als ein anderes Beispiel aus neuester Zeit ist die astronomische Voraussagung zu erwähnen, die *Einstein* aufgrund seiner Gravitationstheorie gemacht hat. Nach der Einsteinschen Theorie muß ein Lichtstrahl, da er Energie besitzt, auch die Eigenschaft der Schwere haben, er muß also von einem schweren Körper, z. B. der Sonne, angezogen werden. Demnach wird ein Lichtstrahl, der von einem Fixstern zur Erde kommt, durch die Sonne von der geradlinigen Bahn abgelenkt. Dieser Effekt ist allerdings so gering, daß er selbst mit feinen Apparaten nur dann zu merken ist, wenn die Bahn des Lichtstrahls nahe an der Sonne vorbeiführt, das heißt, wenn der Stern von der Erde aus in nahezu derselben Richtung liegt, wie die Sonne. Nun können aber solche Sterne für gewöhnlich nicht beobachtet werden, da sie gegenüber dem Sonnenlicht gänzlich zurücktreten; man kann daher die Beobachtung des Ablenkungseffektes nur dann ausführen, wenn eine Verfinsterung der Sonne stattfindet. Es wurden nun vor kurzem bei einer Sonnenfinsternis an verschiedenen Orten Beobachtungen über die Stellung der Sterne in der Nähe der Sonne angestellt, und die Einsteinsche Behauptung hat sich innerhalb der vorausbestimmten Genauigkeitsgrenzen bewährt.

Eine zweite Art der Beziehung zwischen Theorie und Experiment liegt da vor, wo eine Theorie sich auf schon vorhandene Experimente stützt oder durch sie widerlegt wird. So wurden von de Sitter die vorhandenen Beobachtungen an den Doppelsternen zur Widerlegung der Hypothese von Ritz benutzt, wonach sich bei der Lichtausbreitung die Bewegungsgeschwindigkeit des leuchtenden Körpers zu der Fortpflanzungsgeschwindigkeit des Lichtes addiert.

Das Einsteinsche Gravitationsgesetz erhält eine wesentliche Stütze dadurch, daß sich aus ihm als Konsequenz die fortschreitende Bewegung des Merkurperihels ergibt, welche, wie ich schon erwähnt habe, durch astronomische Beobachtung längst bekannt war, aber bisher nur durch eigens eingeführte Hypothesen erklärt werden konnte.

Der Versuch von Fizeau, betreffend die Lichtausbreitung in einer bewegten Flüssigkeit, diente zunächst zur Bestätigung einer speziellen Hypothese über die Mitführung des Äthers durch ein bewegtes Medium, welche Fresnel zur Erklärung eines anderen Experimentes aufgestellt hatte. Durch das Relativitätsprinzip der Elektrodynamik erhielt dann die Fresnelsche Annahme eine überraschend einfache Deutung, und so wurde der Fizeausche Versuch zu einem wichtigen Argument für die Relativitätstheorie.

Der dritte Fall besteht darin, daß eine Theorie eigens dazu gemacht wird, um gewisse experimentell gefundene wichtige Tatsachen zu erklären. Ein solcher Fall tritt insbesondere dann ein, wenn eine Erfahrungstatsache einen so prinzipiellen Charakter trägt, daß sie dem theoretischen Forscher als Fingerzeig für den Ansatz einer Theorie erscheint. Hierfür sind vor allem zwei große Beispiele anzuführen.

Bei der genaueren experimentellen Untersuchung der magnetischen Wirkung des elektrischen Stromes durch Weber und Kohlrausch ergab sich, daß die in dem Biot-Savartschen Gesetze auftretende Konstante (welche gleich dem Verhältnis der Maßzahlen für die Stromstärke in elektrostatischen und elektromagnetischen Einheiten ist) die Größe der Lichtgeschwindigkeit besitzt. Diese Tatsache erschien Maxwell so bedeutsam, daß er sie zum Ausgangspunkt nahm für seine Lichttheorie, durch welche er die Lichtwellen als einen elektromagnetischen Vorgang erklärte.

Die Tatsache, daß die träge Masse (welche das Verhältnis von Kraft und Beschleunigung bestimmt) proportional ist zu der schweren Masse (von der das Gewicht eines Körpers abhängt), war bereits Newton bekannt. Mit Hilfe von Pendelschwingungen stellten Newton und Bessel fest, daß das Verhältnis der trägen zur schweren Masse unabhängig ist von dem Material und dem physikalischen Zustande des Körpers. Genauer wurde dies in neuerer Zeit durch Eötvös nachgewiesen, der die resultierende Richtung aus der Anziehungskraft der Erde und der durch ihre Drehung verursachten Zentrifugalkraft mittels der Torsionswaage für verschiedene Stoffe (Kork, Glas, Antimon) bestimmte; es zeigte sich dabei eine sehr genaue Übereinstimmung.

Während man sich nun seit langem gewöhnt hatte, die Proportionalität (bzw. Gleichheit) von träger und schwerer Masse als etwas Selbstverständliches zu betrachten, erblickte Einstein in dieser Tatsache ein Anzeichen dafür, daß Schwere und Trägheit aus einem gemeinsamen Prinzip zu erklären seien, und dies war für ihn der Leitgedanke bei der Aufstellung der allgemeinen Relativitätstheorie.

Ein ferneres wichtiges Beispiel aus der neueren Physik, bei dem es sich um eine Hypothese von speziellerem Charakter handelt, ist die Erklärung der Spektren durch die *Bohrsche Atomtheorie*. Die Spektren der verschiedenen Stoffe waren mit äußerster Genauigkeit untersucht und ihre eigenartige Struktur forderte zu dem Versuch einer theoretischen Erklärung heraus. Man stand aber dieser Aufgabe lange ratlos gegenüber, bis der Quantenansatz von Bohr mit einem Male Licht brachte. Dieser Ansatz stützte sich auf den Gedanken von Planck, daß die Energie (genauer die Wirkung) aus diskreten Quanten besteht, und auf das Gesetz von Einstein, wonach sich die Frequenz ν der Strahlung, welche von einem Elektron bei einem Energieverlust $(E_1 - E_2)$ ausgesandt wird, aus der (erst von Bohr aufgestellten) Gleichung

$$E_1 - E_2 = h \cdot \nu$$

bestimmt (h bedeutet das Plancksche Wirkungsquantum).* Auf Grund dieser Sätze der Quantentheorie gab Bohr eine Erklärung der Spektren, indem er die Hypothese aufstellte, daß im Atom sich Elektronen um einen Kern bewegen, wobei nur gewisse diskrete Umlaufsbahnen möglich sind, und daß die Ausstrahlung zustandekommt, indem ein Elektron von einer Bahn auf eine andere springt und dabei Energie verliert. Gemäß dem Einsteinschen Quantengesetz hängt dabei die Frequenz der Strahlung von der Anfangs- und der Endbahn, also (da jede Bahn in der diskreten Schar durch eine ganze Zahl charakterisiert ist) von zwei ganzen Zahlen ab. Hierdurch erklärt sich die Verteilung der charakteristischen Linien im Spektrum, insbesondere die Gruppierung der Linien in Serien. Nämlich die Zahl der Anfangsbahn bestimmt die Serie, die der Endbahn die einzelne Linie in der Serie.

Der vierte Fall des Verhältnisses von Theorie und Experiment ist dadurch gekennzeichnet, daß durch eine experimentelle Entdeckung dem physikalischen Forschen ein ganz neues Gebiet erschloßen wird. Hier hat also das Experiment die führende Rolle. Ein Beispiel dieser Art ist der *Oerstedsche Versuch*, mit welchem die Ablenkung der Magnetnadel durch den elektrischen Strom festgestellt wurde. Hier zeigte sich zum ersten Mal

* Die Gleichung selbst hat Bohr erst aufgestellt.

ein Zusammenhang zwischen Elektrizität und Magnetismus, die vordem als ganz getrennte Gebiete der Physik erschienen. Der Zusammenhang wurde dann immer weiter verfolgt, insbesondere durch Faraday, der auch nachwies, daß die Wirkung zwischen Strom und Magnet eine gegenseitige ist. Diese Entwicklung hat soweit geführt, daß in der modernen Elektrodynamik die elektrische und die magnetische Feldstärke zu einer untrennbaren Einheit in Form eines Sechservektors zusammengefaßt werden.

Als ferneres Beispiel ist die *Entdeckung von Röntgen* zu erwähnen, der die nach ihm benannten Strahlen fand, eine neue Art von Strahlung, mit besonderen Eigenschaften. An diese Auffindung der Röntgenstrahlen hat eine fruchtbare Entwicklung physikalischer Erkenntnis angeknüpft. Eine andere bahnbrechende Entdeckung aus neuerer Zeit ist die von Becquerel, welcher bemerkte, daß von dem Stoffe Uran Strahlen ausgesandt werden, welche die Luft leitend machen. Diese Entdeckung gab den Anstoß zu weiteren experimentellen Untersuchungen, durch welche die Physik zu einem ganz neuen Kreis von Erfahrungen, nämlich zur Erkenntnis der Erscheinungen der Radioaktivität gelangte.

Ich habe hier nur eine kleine Auswahl von Beispielen für die vier Fälle gegeben. Natürlich liefert die ganze Physik fortwährend solche Beispiele. Durch die Betrachtung dieser verschiedenen Fälle werden uns zwei Umstände besonders deutlich vor Augen geführt, die in enger Beziehung stehen zu dem, was wir in betreff der beiden ersten Punkte unserer Erörterung der physikalischen Methode gefunden haben: nämlich einerseits, daß der Bereich der zu erklärenden Tatsachen für die Physik nicht von vornherein gegeben ist, sondern durch die experimentellen Entdeckungen beständig erweitert wird—was ja der Grund ist, weshalb die aufgestellten Gesetze immer wieder geändert werden müssen—und andererseits, daß die Bedeutung eines experimentellen Ergebnisses nicht ohne weiteres zutage liegt, sondern sich erst im Zusammenhang der theoretischen Gedankenentwicklung herausstellt, was seine Erklärung findet in der systematischen Tendenz der physikalischen Wissenschaft.

DRITTER TEIL:

FRAGEN PHILOSOPHISCHEN CHARAKTERS

7. KAPITEL: PHYSIKALISCHE GESETZLICHKEIT UND KAUSALITÄT

Nachdem wir nun die primitive Auffassung über das mathematische Denken sowie über die physikalische Forschung geläutert haben, wollen wir von dem gewonnenen Standpunkt einige bestimmte, mehr philosophische Fragen untersuchen. Als erste Frage will ich behandeln: Physikalische Gesetzlichkeit und Kausalität.

Indem wir die Art der Darstellung der Naturvorgänge zugrunde legen, wie sie in der mathematischen Physik stattfindet, können wir sagen: das Naturgeschehen in Vergangenheit, Gegenwart und Zukunft würde vollständig bekannt sein, wenn gewisse Zustandsgrößen $u_1, \ldots, u_n$ als Funktionen von Ort und Zeit bekannt wären. Das heißt, das Geschehen ist bestimmt durch die räumlich-zeitliche Wertverteilung

$$\begin{aligned} u_1 &= f_1(x,y,z,t) \\ \vdots &= \quad \vdots \\ u_n &= f_n(x,y,z,t) \end{aligned} \tag{1}$$

wobei x, y, z, t die Raum- und Zeitkoordinaten sind, und $u_1 \ldots, u_n$ gewisse Größen, durch welche der physikalische Zustand an einer Raum-Zeitstelle charakterisiert ist.

Beispiele für Systeme von Zustandsgrößen haben wir in speziellen physikalischen Theorien. In der *Hydrodynamik* betrachtet man den Zustand als vollständig bestimmt durch die Verteilung der materiellen Geschwindigkeiten (bzw. deren Komponenten) und der Dichte; in der reinen Feldtheorie des *Elektromagnetismus* bestimmt sich alles durch die vier Komponenten des Viererpotentials, und in der neuen *Feldtheorie*, welche die Gravitation mit umfaßt, bilden die 10 Gravitationspotentiale (die Koeffizienten in dem Ausdruck für das Quadrat des raum-zeitlichen Linienelements) zusammen mit den vier Komponenten des elektrischen Potentials die Bestimmungsgrößen des physikalischen Zustandes. (Die Einführung der Zustandsgrößen muß so geschehen, daß dabei eine Anweisung gegeben wird, wie die Werte der Zustandsgrößen durch Beobachtungen zu ermitteln sind. Zu beachten ist ferner, daß die Auswahl der bestimmenden Zustandsgrößen immer in einem gewissen Masse willkürlich bleibt.)

Gesetzt nun, wir hätten eine alles umfassende physikalische Theorie und die Funktionen, welche die Werte der Zustandsgrößen in ihrer Abhängigkeit von Ort und Zeit darstellen, wären bekannt, so hätten wir damit alle mathematischen Daten zu einer vollständigen Naturbeschreibung. In welcher Weise können uns aber diese Funktionen bekannt sein? Wie ist hier Bekanntheit zu verstehen? Um das Grundsätzliche dieser Frage scharf aufzufassen, wollen wir soviel wie möglich das Problem vereinfachen. Denken wir uns das Weltgeschehen schematisch repräsentiert durch ein Würfelspiel. Es fallen dann die räumlichen Ortsbestimmungen ganz weg und zeitlich kommen nur diskrete Momente in Betracht; der momentane Zustand wird durch eine einzige Größe charakterisiert, nämlich durch die jeweils auf der obersten Würfelfläche befindliche Ziffer.

Das Weltgeschehen würde also dargestellt durch eine Funktion, deren Argument nur ganzzahlige Werte durchläuft (eine zahlentheoretische Funktion) und deren Werte

$$f(1), f(2), f(3), \ldots$$

auf die Zahlen von 1 bis 6 beschränkt sind. Nehmen wir die Funktionswerte $f(1), f(2), f(3), \ldots$ als Ziffern einer hexadischen Bruchentwicklung, so wird das schematisch vereinfachte Weltgeschehen durch einen unendlichen Bruch, also eine einzige (reelle) Zahl ausgedrückt. Jetzt lautet unsere Frage: Wie kann ein unendlicher hexadischer Bruch bekannt sein? Eine Aufzählung der aufeinander folgenden Ziffern kann nicht ausgeführt werden, weil die Zahl der Stellen unendlich ist. Die Aufzählung wäre mit dem Weltgeschehen selbst identisch. Der hexadische Bruch kann also nur in der Weise mitgeteilt werden, daß ein Gesetz für seine Entwicklung angegeben wird. Dies ist eine dem Mathematiker geläufige Tatsache, daß eine unendliche Reihe nicht anders als durch ein Rechengesetz gegeben werden kann. Ein besonders einfacher Fall eines solchen Entwicklungsgesetzes würde der sein, daß ein und dieselbe Aufeinanderfolge von Ziffern immer wiederkehrt. Das Weltgeschehen wäre dann periodisch, also eine Welle.

Sehen wir nun von der schematischen Vereinfachung ab, so können wir aus unserer Überlegung jedenfalls dies entnehmen, daß zur grundsätzlichen Möglichkeit eines vollständigen Kennens der Natur bereits eine gewisse Gesetzmäßigkeit erfordert wird. Dieses Ergebnis besteht nicht nur unter der Voraussetzung der Unendlichkeit der Welt, sondern es bleibt auch dann noch gültig, wenn man nur eine endliche Mannigfaltigkeit von Bestimmungen in der Welt annimmt. Nämlich auch in diesem Falle wäre eine vollständige Aufzählung aller Daten unmöglich. Denn eine solche Aufzählung würde ja eine größere Mannigfaltigkeit enthalten als die Welt selbst, indem zu den Bestimmungsstücken des Wirklichen noch die Hilfsmittel der Bezeichnung hinzukämen. Das Weltgeschehen selbst ist ja schon die einfachst mögliche Tabelle. Eine vollständige Aufzählung aller Bestimmungsstücke des Wirklichen könnte daher nicht in dem Weltgeschehen (als ein einzelner Vorgang) enthalten sein. Auf jeden Fall ist also eine vollständige Kenntnis der Wirklichkeit nur dadurch möglich, daß irgendeine Art von *Gesetzlichkeit* besteht.

Wenn demnach Kirchhoff erklärt, die Aufgabe der Physik bestehe in der Beschreibung der Wirklichkeit, so kann dieses Beschreiben nicht anders geschehen als durch Auffindung von Gesetzen.

In welcher Beziehung steht nun aber die Gesetzlichkeit des Geschehens zur *Kausalität*? Jedenfalls ist in dem Gedanken der Kausalität enthalten, daß das Naturgeschehen sich nach Gesetzen vollzieht.* Aber das allein ist zur Kausalität nicht ausreichend. Das Prinzip der Kausalität verlangt insbesondere, daß unter gleichen Umständen stets Gleiches geschieht. Diese Forderung hat aber mit derjenigen der Gesetzlichkeit gar nichts zu tun. Wir können uns das leicht mathematisch klar machen. Daß unter gleichen Umständen Gleiches geschieht, ist eine Bedingung, welche den Funktionen

$$f_1(x,y,z,t),\ \ldots,\ f_n(x,y,z,t)$$

auferlegt wird, ohne daß damit gesagt ist, daß die Funktionen durch ein Gesetz bestimmt sind. Die Bedingung ist z. B. sicher erfüllt, wenn niemals gleiche Umstände sich wiederholen, obwohl damit noch keinerlei Gesetzlichkeit gegeben ist.

Auch umgekehrt folgt aus dem Bestehen einer Gesetzlichkeit noch keineswegs, daß unter gleichen Umständen Gleiches geschieht. Ließe sich beispielsweise das Weltgeschehen darstellen in der Form

$$\begin{array}{ccc} u_1 & = & g_1(x,y,z)\cdot t\cdot \sin t \\ u_2 & = & g_2(x,y,z)\cdot t\cdot \sin t \\ \vdots & \vdots & \vdots \\ u_n & = & g_n(x,y,z)\cdot t\cdot \sin t \end{array}$$

(wobei $g_1(x,y,z),\ \ldots,\ g_n(x,y,z)$ bestimmte mathematische Funktionsausdrücke bedeuten sollen), so wäre das Geschehen wohl ein gesetzliches, aber es würde unter gleichen Umständen nicht immer das Gleiche eintreten. Denn zu den Zeiten

$$t = 0,\ t = \pi,\ t = 2\pi, \ldots$$

würden die gleichen Umstände vorliegen (weil ja die Zustandsgrößen an jedem Ort sämtlich gleich Null wären); es würden aber nicht die gleichen Zustände nachfolgen (wie sich aus dem Auftreten des Faktors t ergibt). Es sind also zwei ganz verschiedene Voraussetzungen, daß unter gleichen Umstände dasselbe geschieht und daß das Naturgeschehen gesetzlich ist. Wir

* Die Kantische Formulierung des Kausalitätsprinzips lautet: “Alles was geschieht (anhebt zu sein) setzt etwas voraus, worauf es nach einer Regel folgt.”

wollen zunächst die erste dieser beiden in dem Kausalitätsprinzip enthaltenen Forderungen näher verfolgen und sie auf mathematische Form bringen.

Da das Geschehen in der Aufeinanderfolge von Zuständen besteht und da ferner die zeitliche Änderung eines Zustandes mathematisch dargestellt wird durch die nach der Zeit t genommenen Differentialquotienten der Zustandsgrößen

$$\frac{\partial u_1}{\partial t}, \ldots, \frac{\partial u_n}{\partial t}$$

so kommt die Behauptung, daß unter gleichen Umständen Gleiches geschieht, darauf hinaus, daß die Wertverteilung der partiellen Ableitungen

$$\frac{\partial u_1}{\partial t}, \ldots, \frac{\partial u_n}{\partial t}$$

(welche Funktionen von x, y, z, t sind) in eindeutiger Weise von den Umständen abhängt. Hierbei muß noch näher bestimmt werden, was mit den "Umständen" gemeint ist.

Versteht man darunter den zu einer Zeit bestehenden Gesamtzustand, so besagt unser Satz, daß in jedem Zeitpunkt die Zustandsänderung eindeutig aus dem Gesamtzustand hervorgeht; und die mathematische Formulierung wird dann durch ein Gleichungssystem der Form

$$\begin{aligned} \frac{\partial u_1}{\partial t} &= \Phi_1(u_1, \ldots, u_n, x, y, z) \\ \vdots \quad & \quad \vdots \\ \vdots \quad & \quad \vdots \\ \frac{\partial u_n}{\partial t} &= \Phi_n(u_1, \ldots, u_n, x, y, z) \end{aligned} \tag{2}$$

gegeben, wobei die Abhängigkeit der $\Phi_1, \ldots, \Phi_n$ von den Argumenten so allgemein aufzufassen ist, daß dabei sowohl Funktionen wie auch Ableitungen nach den Raumkoordinaten auftreten können.

Ein Beispiel eines solchen Gleichungssystems bilden die Formeln, welche für ein materielles Kontinuum bei alleiniger Wirkung von Newtonschen Gravitationskräften gelten. Hier hat man als Zustandsgrößen die Komponenten der Geschwindigkeit u, v, w und die Dichte ρ. Die Gleichungen lauten:

$$\frac{\partial u}{\partial t} = \iiint \frac{\rho(a,b,c,t)(a-x)}{\left((a-x)^2+(b-y)^2+(c-z)^2\right)^{3/2}}\, da\, db\, dc$$
$$\frac{\partial v}{\partial t} = \iiint \frac{\rho(a,b,c,t)(b-y)}{\left((a-x)^2+(b-y)^2+(c-z)^2\right)^{3/2}}\, da\, db\, dc$$
$$\frac{\partial w}{\partial t} = \iiint \frac{\rho(a,b,c,t)(c-z)}{\left((a-x)^2+(b-y)^2+(c-z)^2\right)^{3/2}}\, da\, db\, dc$$
$$\frac{\partial \rho}{\partial t} = -\Big(\frac{\partial(\rho\cdot u)}{\partial x} + \frac{\partial(\rho\cdot v)}{\partial y} + \frac{\partial(\rho\cdot w)}{\partial z}\Big).$$

Die wesentliche Eigenschaft der Gleichungen (2), durch welche die Voraussetzung gleichen Geschehens unter gleichen Umständen zum Ausdruck kommt, besteht darin, daß die Zeitkoordinate t nicht explizite auftritt. (Bekanntlich beruht in der Mechanik auf dieser Eigenschaft der Bewegungsgleichungen die Gültigkeit des Energiesatzes.) Tatsächlich macht man nun aber in der neuen Physik eine schärfere Voraussetzung. Gemäß dem *Prinzip der Nahewirkung* nimmt man an, daß die Zustandsänderung an einer Stelle bestimmt ist durch die Wertverteilung der Zustandsgrößen in unmittelbarer Nachbarschaft dieser Stelle; d. h. man betrachtet als die Umstände, von denen die Änderungen

$$\frac{\partial u_1}{\partial t}, \ldots, \frac{\partial u_n}{\partial t}$$

abhängen, nicht den Gesamtzustand zur Zeit t, sondern den (zu dieser Zeit bestehenden) Zustand in der Umgebung der betreffenden Raumstelle x, y, z. Nach dieser Auffassung hat man Gleichungen der Form

$$\begin{aligned} \frac{\partial u_1}{\partial t} &= F_1\Big(u_1, \ldots, u_n, \frac{\partial u_1}{\partial x}, \frac{\partial u_1}{\partial y}, \ldots, \frac{\partial u_n}{\partial z}, \ldots\Big) \\ &\;\vdots \\ &\;\vdots \\ \frac{\partial u_n}{\partial t} &= F_n\Big(u_1, \ldots, u_n, \frac{\partial u_1}{\partial x}, \frac{\partial u_1}{\partial y}, \ldots, \frac{\partial u_n}{\partial z}, \ldots\Big) \end{aligned} \tag{3}$$

anzusetzen, wobei $F_1, \ldots, F_n$ Funktionen im gewöhlichen Sinne bedeuten, als deren Argumente die Zustandsgrößen $u_1, \ldots, u_n$ und ihre räumlichen Ableitungen (eventuell auch von höherer als erster Ordnung) auftreten.

In diesen partiellen Differentialgleichungen kommt insbesondere zum Ausdruck:

1) die Unabhängigkeit der Naturgesetze von Ort und Zeit, indem die Koordinaten x, y, z, t nicht explizite vorkommen;

2) die Stetigkeit des Geschehens, im Sinne der Nahewirkung, indem für die Veränderung an einem Orte nicht nur die Werte der Zustandsgrößen an dieser Stelle, sondern auch die räumlichen Ableitungen maßgebend sind.

Damit man die Einschränkung, welche in der Form der Gleichungen (3) enthalten ist, nicht überschätzt, ist es nützlich, sich folgendes klarzumachen: Für die Gültigkeit eines Gleichungssystems von der Form (3) würde hinreichend sein, daß zunächst die Zustandsgrößen nach den Koordinaten differenzierbar sind und daß überdies niemals an zwei verschiedenen Raum-Zeitstellen derselbe Zustand herrscht. Denn unter dieser Voraussetzung würden ja die Werte der Koordinaten x, y, z, t in eindeutiger Weise durch den Zustand, d. h. durch die Werte der $u_1, \ldots, u_n$ und ihre räumlichen Ableitungen bestimmt sein, und die zeitlichen Differentialquotienten

$$\frac{\partial u_1}{\partial t}, \ldots, \frac{\partial u_n}{\partial t},$$

welche ja, gemäß den Gleichungen

$$u_1 = f_1(x, y, z, t), \ \ldots, \ u_n = f_n(x, y, z, t),$$

Funktionen von x, y, z, t sind, würden sich also als Funktionen der $u_1, \ldots, u_n$ und ihrer räumlichen Ableitungen darstellen, so wie es die Gleichungen (3) verlangen. Allerdings wird die hier betrachtete Voraussetzung in Wirklichkeit schwerlich erfüllt sein. Es braucht ja z. B. nur eine kleine Menge ruhender, homogener Flüssigkeit zu existieren; dann sind in einem Raum-Zeitgebiet die Werte der Zustandsgrößen konstant.

Durch den Ansatz der Gleichungen (2) haben wir mathematisch die Forderung zun Ausdruck gebracht, daß unter gleichen Umständen Gleiches geschieht. Nun müssen wir den Gedanken der *Gesetzlichkeit* hinzunehmen. Die Annahme einer Gesetzlichkeit besagt zunächst, daß in den Gleichungen (3) die $F_1, \ldots, F_n$ nicht nur als Funktionsschemata, sondern als bestimmte, durch ein mathematisches Gesetz angebbare Funktionen auftreten. Nun ist der Begriff eines mathematischen Gesetzes außerordentlich allgemein, und auch wenn wir uns auf stetige Funktionen beschränken, so können die Funktionsgesetze noch höchst kompliziert sein. Wenn wir aber von der Gesetzlichkeit der Natur sprechen, so liegt darin der Gedanke, daß die Abhängigkeit des Geschehens von den Umständen durch einfache Gesetze bestimmt wird. Die Idee der *Einfachheit* tritt also hier als wesentliches Moment auf.

Bei diesem Punkte zeigt der mathematische Formalismus dadurch seine große Bedeutsamkeit, daß er es erst ermöglicht, den zunächst vagen Begriff der Einfachheit zu präzisieren. Die einfachsten Funktionen sind die ganzen rationalen Funktionen von niedrigem Grade (insbesondere die linearen und quadratische Funktionen). Hat man es mit Differentialausdrücken zu tun, so besagt die Forderung der Einfachheit ferner, daß nur Ableitungen von niedriger Ordnung auftreten dürfen. Beim Ansetzen der Differentialgleichungen (3) wird man also die Funktionen $F_1, \ldots, F_n$ als ganz rationale Funktionen niedrigen Grades der $u_1, \ldots, u_n$ und ihrer Ableitungen nach x, y, z zu bestimmen suchen, und zwar wird man von den partiellen Ableitungen nur die niedrigsten, wenn möglich nur die ersten und zweiten Ableitungen, in Betracht ziehen.

In einem derartigen Ansatz hat man aber noch sehr viele Konstanten zur Verfügung, und es bleibt daher für die theoretische Zusammenfassung der Erfahrungstatsachen noch ein großes Maß von Willkür. Es spricht hierbei der Umstand mit, daß die Anweisung, welche wir durch die Experimente erhalten, auf jeder Entwicklungsstufe der Wissenschaft insofern unvollkommen ist, als die physikalischen Messungen sich ja niemals absolut genau und auch nur für beschränkte Wertgebiete der veränderlichen Größen ausführen lassen. Wir können also aufgrund von Beobachtungen eine Funktion immer nur insoweit bestimmen, daß ein sie umgebender Streifen abgegrenzt wird, und dies auch nur in einem beschränkten Wertbereich der Argumente, so daß für die Wahl des mathematischen Ausdrucks stets noch ein beträchtlicher Spielraum bleibt.

Man wird daher veranlaßt, das Prinzip der Einfachheit in noch stärkerem Masse in Anspruch zu nehmen. Und dies tut man, indem man Vorschriften prinzipieller Art für die Differentialgleichungen aufstellt. Insbesondere sind folgende zwei Vorschriften zu erwähnen:

1) Die Differentialgleichungen sollen aus einem Variationsproblem entspringen (Minimaleigenschaft der Naturprozesse).

2) Die Differentialgleichungen sollen bei beliebigen Koordinatentransformationen (sofern diese nur den Bedingungen der Differenzierbarkeit und der eindeutigen Umkehrbarkeit genügen) ihre Form behalten (Forderung der Invarianz).

Diese beide Vorschriften erhalten erst eigentlich ihre Bedeutung, wenn sie im Sinne des Prinzips der Einfachheit angewandt werden. Man kann nämlich, *cum grano salis*, jedes System von Differentialgleichungen auf ein Variationsprinzip zurückführen; nur wird dann im allgemeinen der Ausdruck dieses Variationsprinzip sehr kompliziert. Ebenso läßt sich auch jedes Differentialgleichungssytem (durch formale Einführung neuer Funktionen) auf ein invariante Form bringen; nur führt ein solches Verfahren für gewöhnlich zu einer höchst komplizierten Darstellung. Für sich genommen ergeben also die zwei genannten mathematischen Forderungen keine Beschränkung in der Auswahl der Gesetze. Vom Standpunkt der Einfachheit haben sie aber die Bedeutung, daß unter den Differentialgleichungssystemen, welche in dem vorhin erläuterten Sinne einfach sind, diejenigen ausgewählt werden sollen, bei denen auch noch die Darstellung durch ein invariantes Variationsproblem in einfacher Form möglich ist.

Durch Bedingungen solcher Art wird die Mannigfaltigkeit der in Betracht kommenden Gesetze sehr beschränkt, und man gewinnt so aus dem Prinzip der Einfachheit einen weitgehenden Anhalt für die Bildung physikalischer Theorien. Besonders in der modernen Physik erweist sich der Gesichtspunkt der formalen mathematischen Einfachheit als fruchtbar. Die Grundgleichungen der Maxwellschen Theorie erhalten in der Formalsprache der Viereranalysis eine überraschend einfache Form. Sie bieten sich als die einfachsten überhaupt in Betracht kommenden Differentialgesetze dar, sobald einmal die Begriffsbildungen des Viererpotentials, des elektromagnetischen Sechservektors und der Viererdichte zugrunde gelegt sind.

Während diese große mathematische Einfachheit der elektrodynamischen Grundgesetze sich als eine unerwartete Entdeckung ergab, ist in der neuen Gravitationstheorie von Einstein ganz bewußt und planmäßig das Prinzip der formalen Einfachheit zum leitenden Gedanken für die Aufstellung seiner Feldgleichungen der Gravitation gewählt worden, und dieses Verfahren hat zum größten Erfolge geführt, indem es nicht nur gelang, das alte Gravitationsgesetz von einem allgemeinen theoretischen Standpunkt aus (als Näherungsgesetz) abzuleiten, sondern überdies auf einem systematisch festbestimmten Wege eine Korrektur des Newtonschen Gesetzes gefunden wurde, aufgrund deren das bisher unerklärte (d. h. nur durch künstliche Zusatzhypothesen erklärbare) Fortschreiten des Perihels in der Bahn des Merkur sich als notwendige Konsequenz ergibt.

Noch einen weiteren Gesichtspunkt für die Beurteilung der Einfachheit von physikalischen Ansätzen gewinnen wir, wenn wir uns klar machen, daß wir bisher bei unserer Betrachtung eine Fiktion angewandt haben, indem wir von der Vorstellung einer alles umfassenden physikalischen Theorie ausgegangen sind, deren Grundgesetze durch das Gleichungssystem (3) formuliert sein sollten. Eine solche endgültige Theorie ist ja in Wirklichkeit nicht vorhanden, sondern schwebt nur als Ideal der physikalischen Forschung vor. Die tatsächlich aufgestellten physikalischen Theorien sind immer nur auf bestimmte Klassen von Naturvorgängen anwendbar.

Hiermit ergibt sich ein neuer Maßstab für die Einfachheit einer Theorie. Denn eine Theorie entspricht um so mehr der Forderung der Einfachheit, je größer der Bereich der Tatsachen ist, welche sie in ein Gesetz zusammenfaßt. Wenn für spezielle, experimentell verfolgbare Abhängigkeiten durch Probieren einfache mathematische Formeln gewonnen werden (wie man es z. B. bei der Verbesserung der van der Waalsschen Zustandsgleichung gemacht hat), so ist damit in Hinsicht auf die Einfachheit der Naturerklärung nicht viel erreicht, und man hat keine Gewähr, das Richtige zu treffen. Werden dagegen durch eine Theorie große Klassen von Tatsachen unter einem Gesetz vereinigt, so erblicken wir darin ein Anzeichen dafür, daß wir der wahren

Gesetzlichkeit in dem betreffenden Tatsachengebiet auf die Spur gekommen sind.

Ein glänzendes Beispiel hierfür bietet die Maxwellsche Theorie des Elektromagnetismus. Diese schließt als Folgerung in sich: das Coulombsche Gesetz (Gesetz für die Anziehung und Abstoßung elektrischer Ladungen), das Biot-Savartsche Gesetz (Gesetz für die Ablenkung einer Magnetnadel durch den elektrischen Strom) und das Faradaysche Induktionsgesetz (Gesetz der Abhängigkeit zwischen veränderlichem Magnetfeld und elektrischem Strom); ferner erklärt sie die Tatsachen der drahtlosen Telegraphie; die Wellentheorie des Lichtes läßt sich aus ihr ableiten, und die numerische Übereinstimmung der Konstanten des Biot-Savartschen Gesetzes (d. h. des Verhältnisses zwischen dem Werte der Stromstärke in elektrostatischem und elektromagnetischem Maß) mit der Lichtgeschwindigkeit findet durch sie ihre theoretische Deutung.

Der Erfolg des Prinzips der mathematischen Einfachheit in der Physik ist ein erstaunlicher. Wenn man die überraschend einfache Form gewahr wird, welche die Grundgleichungen der Maxwellschen Theorie in der Formelsprache der Viereranalysis annehmen, und wenn man ferner sieht, wie in der Einsteinschen Feldtheorie der Gravitation die Heranziehung der einfachsten Differentialinvarianten die richtige Korrektur des Newtonschen Gravitationsgesetzes liefert, dann kommt man zu dem Eindruck einer prästabilierten Harmonie. Wir stehen da der merkwürdigen Tatsache gegenüber, daß anscheinend die Materie sich ganz und gar dem Formalismus der Mathematik fügt. Es zeigt sich hier ein unvorhergesehener Einklang zwischen Sein und Denken, den wir vorläufig wie ein Wunder hinnehmen müssen.

Auf diese wunderbare Angemessenheit des mathematischen Formalismus zur Darstellung der Naturgesetze sind wir erst von verschiedenen Seiten her gestoßen worden. Die Menschen haben zumeist keine Vorliebe für mathematischen Formalismus; nicht etwa haben sie diesen künstlich in die Naturwissenschaften hineingetragen, sondern sie wurden durch die Nötigung der Tatsachen zu seiner immer weiteren Anwendung gedrängt. (Vielleicht gelingt es einmal, alle physikalischen Theorien zu einer mathematischen Gesamtwissenschaft, nach Art der Geometrie zu vereinigen, bei der auch keine nur erfahrungsmäßig zu bestimmenden Konstanten mehr in den Gesetzten auftreten. Diese Möglichkeit ist neuerdings viel näher gerückt.)

Unsere Betrachtung über die Forderung der Einfachheit bezieht sich auf die Darstellung der Naturgesetzlichkeit durch Differentialgleichungen von der Form (3). Die Einfachheit der *Naturgesetze* bedeutet aber noch keineswegs die Einfachheit des *Naturgeschehens*, oder mathematisch ausgedrückt: Aus der Einfachheit der Differentialgleichungen (3) (bzw. der Funktionen $F_1, \ldots, F_n$) folgt noch keineswegs die Einfachheit der Funktionen $f_1, \ldots, f_n$, welche *Lösungen* der Differentialgleichungen bilden.

Dies zeigt sich schon bei dem Fall der Newtonschen Gravitationstheorie. Hier sind die Differentialgleichungen von einfacher Form; hingegen sind die Lösungen bereits beim Dreikörperproblem sehr kompliziert. Die Bewegungen der Himmelskörper, auf deren direkte Beschreibung die Ptolemäische Astronomie ausging, sind sehr verwickelt. Erst durch den Übergang zu Differentialgleichungen gelangte Newton zu seinem erstaunlich einfachen Gravitationsgesetz.

Wir müssen uns überhaupt klar machen, daß durch die Differentialgleichungen das Weltgeschehen nicht völlig bestimmt ist. Denn in den Lösungen der Differentialgleichungen (3) treten noch willkürliche Funktionen auf. Die Differentialgleichungen enthalten das Gesetz der Veränderung, und zur vollständigen Bestimmung des Geschehens würde noch gehören, daß zu irgendeiner Zeit die Wertverteilung der Funktionen $f_1, \ldots, f_n$ gegeben ist. Es ist daher auch durch die Voraussetzung der Gesetzlichkeit, im Sinne des Kausalitätsprinzips, gar nicht ausgemacht, daß die Zustandsgrößen überhaupt durch ein mathematisches Gesetz als Funktionen von Ort und Zeit bestimmt sind.

Nun ist allerdings aus der Theorie der partiellen Differentialgleichungen bekannt, daß man bei gewissen Klassen von Differentialgleichungen die willkürlichen Funktionen durch Randbedingungen festlegen kann. So wird in der Potentialtheorie unter den Lösungen der Differentialgleichung

$$\frac{\partial^2 u}{\partial x^2} + \frac{\partial^2 u}{\partial y^2} + \frac{\partial^2 u}{\partial z^2} = 4\pi\rho$$

eine spezielle ausgewählt, indem man das Verhalten der Funktion u im Unendlichen vorschreibt. Hierdurch wird der Gedanke nahe gelegt, ob nicht bei den Differentialgleichungen der Physik diejenige Lösung, welche den Ablauf des Naturgeschehens darstellt, durch Randbedingungen bestimmt werden kann. Insbesondere wird man diese Frage mit Bezug auf Einsteins neue Feldtheorie der Gravitation stellen.

Hier steht es nun so, daß hinsichtlich des Raumes von Einstein eine Art von Randbedingung aufgestellt wird. Er macht nämlich die (aus anderweitigen theoretischen Gründen plausible) Annahme, daß die Welt dem Raume nach endlich und geschlossen ist, so daß es für das Räumliche überhaupt kein Unendliches und keine Grenze (keinen Rand) gibt. Durch diese Annahme wird aber noch keineswegs eine Lösung der Feldgleichungen festgelegt. Schon darum nicht, weil ja hinsichtlich der Zeit keine Randbedingung aufgestellt ist.

Aber selbst wenn man sich dazu entschließen wollte, die Welt auch der Zeit nach als endlich und geschlossen, oder aber (was mathematisch auf dasselbe hinauskäme) das Weltgeschehen als streng periodisch anzunehmen, so würde die Festlegung einer Lösung der Feldgleichung damit nicht erreicht werden. Der Grund hierfür liegt darin, daß die Einsteinschen Feldgleichungen nicht den Charakter der Potentialgleichung

$$\frac{\partial^2 u}{\partial x^2} + \frac{\partial^2 u}{\partial y^2} + \frac{\partial^2 u}{\partial z^2} = 0$$

sondern den der Schwingungsgleichung

$$\frac{\partial^2 u}{\partial x^2} + \frac{\partial^2 u}{\partial y^2} + \frac{\partial^2 u}{\partial z^2} - \frac{\partial^2 u}{\partial t^2} = 0$$

besitzen.

Dieser bei der Einsteinschen Theorie, sowie bei allen früheren Theorien sich zeigende Umstand, daß das Naturgeschehen durch die physikalischen Gesetze (Differentialgleichungen und Randbedingungen) noch nicht bestimmt wird, hat eine grundsätzliche philosophische Bedeutung. Könnte man die Funktionen $f_1, \ldots, f_n$ vollständig durch Differentialgleichungen und Randbedingungen festlegen, so hätte man eine Art *Hegelscher Physik*, in der nicht nur bei gegebenen Anfangsbedingungen die nachfolgenden Veränderungen, sondern überhaupt alle Naturbestimmungen insgesamt (z. B. die Anzahl der Sonnenplaneten, die Anzahl der Erdteile) sich durch mathematische Ableitung aus allgemeinen Gesetzten ergeben müßten.

Bisher haben wir zwei Momente des Kausalitätsgedankens betrachtet: einmal die Forderung, daß unter gleichen Umständen Gleiches geschieht, welche wir unter Hinzunahme des Prinzips der Nahewirkung durch den Ansatz der Differentialgleichungen (3) mathematisch zum Ausdruck brachten; und dann die Forderung der Gesetzlichkeit, welche uns zu dem Prinzip der Einfachheit und zur Untersuchung der Kriterien für die Einfachheit führte. Hiermit ist aber das Problem der physikalischen Kausalität noch nicht erschöpft. Denn durch jene beiden Momente ist das Verhältnis von Ursache und Wirkung, so wie wir es aufzufassen pflegen, noch nicht vollständig charakterisiert. Insbesondere kommt ja in den mathematischen Gleichungen gar nicht zum Ausdruck, daß die Ursache das frühere, die Wirkung das spätere ist. Die Berücksichtigung der *Bedeutung des Zeitverhältnisses für die Kausalität* führt uns auf die allgemeine Frage, ob und inwiefern überhaupt eine Zeitrichtung in der Physik ausgezeichnet ist. Wir kommen also auf die Frage der *Reversibilität und Irreversibilität* physikalischer Vorgänge, welche insbesondere seit der Aufstellung des Gesetzes von der Vermehrung der Entropie durch Clausius in der Physik aktuell geworden ist.

Dieses Gesetz wird neuerdings in der kinetischen Gastheorie als ein bloßes Wahrscheinlichkeitsgesetz erklärt, und überhaupt sucht man die Tatsache, daß bei vielen Klassen von Naturvorgängen der zeitliche Ablauf immer nur in einem Sinne (in einer bestimmten Zeitrichtung) beobachtet wird

(z. B. bei akustischen Vorgängen und bei der Strahlung), durch Wahrscheinlichkeitsbetrachtungen verständlich zu machen. Ehe wir daher das Problem der ausgezeichneten Zeitrichtung erörtern können, müssen wir uns erst mit der Bedeutung des Wahrscheinlichkeitsbegriffs für die Physik etwas näher befassen. Wir wenden uns also zu dem Thema: Naturgeschehen und Wahrscheinlichkeit.

8. KAPITEL: NATURGESCHEHEN UND WAHRSCHEINLICHKEIT

Wir haben als eine charakteristische Eigenschaft der durch die Differentialgleichungen formulierten Naturgesetze ihre Unabhängigkeit von Ort und Zeit festgestellt. Diese Unabhängigkeit würde aber gar nicht zur Geltung kommen, wenn sich nicht in der Natur ***Wiederholungen*** und Ähnlichkeiten vorfänden. Ohne diese könnten wir niemals zur Auffindung jener Gesetze gelangen. Die physikalische Forschung geht ja aus von der Aufstellung empirischer Gesetze (wie z. B. der Keplerschen Gesetze der Planetenbewegung), welche an Hand von Beobachtungen gefunden werden, und für diese Beobachtungen bildet die Wiederkehr gleichartiger Verhältnisse (also beim Fall der Planetenbewegung die Wiederkehr des gleichartigen Umlaufes bei den verschiedenen Planeten) ein wesentliches Erfordernis.

Damit wir also die physikalischen Differentialgesetze entdecken können, müssen die in Natur verwirklichten Lösungen der Differentialgleichungen so beschaffen sein, daß darin vielfache Wiederholungen gleicher Umstände vorkommen. Nun gibt es aber in der Natur noch eine besondere Art von Wiederholungen, welche hinausgeht über das, was zur Erkennbarkeit der Naturgesetzlichkeit erforderlich wäre. Wir finden nicht nur eine Wiederkehr gleichartiger Umstände, sondern auch eine Wiederholung derselben Größenbestimmungen. Es gibt eine Unzahl annähernd gleicher Exemplare von Gegenständen und Vorgängen; die Fixsterne haben alle ungefähr dieselbe Größe und auch ungefähr die gleichen Geschwindigkeiten; die Körner des Sandes sind einander nahezu gleich; die Körper bestehen aus einer großen Zahl von nahezu gleichen Molekülen, die sich bei einer bestimmten Stoffart alle in gleicher Weise in Atome zerlegen; die kleinsten Bestandteile der Materie, die Elektronen, haben wiederum alle dieselbe Ausdehnung und gleiche elektrische Ladung; ferner die Lichtwellen bestehen aus einer ungeheuren Zahl von ungefähr gleich schnellen und nahezu gleichweiten Schwingungen; und neuerdings hat sich für die Energie auch die Annahme als notwendig gezeigt, daß sie aus abgesonderten Teilen besteht, die alle von gleichem Betrage sind.

Diese Eigenschaft der Natur ist unabhängig von der Gültigkeit und auch von den Bedingungen der Auffindbarkeit der Differentialgleichungen. Es liegt also darin eine ganz besondere, merkwürdige Erfahrungstatsache, von so prinzipiellem Charakter, daß sie zu ganz neuen Methoden und Theorien in der Physik Anlaß gibt, welche für die theoretische Beherrschung der Erscheinungen unerläßlich sind.

Eine der allereinfachsten Tatsachen, der wir vom Standpunkt der Differentialgleichungen ganz ratlos gegenüberstehen, ist folgende: wenn ich einen Liter weißen Sand mit einem Liter schwarzen Sand mische und greife dann aus der Mischung mit der Hand einen Teil heraus, so fasse ich ungefähr ebensoviel weißen wie schwarzen Sand. Wir sehen schon, daß wir es hier mit einer Massenerscheinung zu tun haben, bei welcher der Zufall eine Rolle spielt. Das *Gesetz des Zufalls* müssen wir also studieren, um diese Tatsache zu erklären.

Zum Zweck der mathematischen Behandlung werden wir zunächst die Aufgabe etwas präzisieren; der Einfachheit halber wollen wir annehmen, daß die weißen und die schwarzen Sandkörner in gleicher Zahl (n) vorhanden sind; h bedeute das Verhältnis der Anzahl sämtlicher zur Anzahl der herausgegriffenen Sandkörner (also ungenau ausgedrückt: das Volumenverhältnis zwischen der ganzen Menge und der herausgenommenen Menge Sand). Es sei also, wenn $2m$ die Anzahl der herausgegriffenen Sandkörner bezeichnet,

$$n = h \cdot m \quad .$$

(Da h als gegeben anzusehen ist, so sind n und m gegenseitig durch einander bestimmt.)

Nun soll durch das Gesetz des Zufalls, d. h. durch Wahrscheinlichkeit erklärt werden, daß sich unter den $2m$ herausgegriffenen Sandkörnern nahezu m weiße und m schwarze befinden (wobei die Bestimmung "nahezu" noch einer Präzisierung bedarf). Dazu berechnet man zuerst die Anzahl aller Arten, wie aus $2n$ Körnern $2m$ herausgenommen werden können; diese ist

$$N = \frac{(2n)!}{(2m)!\,(2n-2m)!} \quad .$$

Die Anzahl der Möglichkeiten, wie $(m+k)$ weiße und $(m-k)$ schwarze Körner herausgegriffen werden können, ist

$$N_k = \frac{n!}{(m+k)!\,(n-m-k)!} \cdot \frac{n!}{(m-k)!\,(n-m+k)!} \quad .$$

(Und dieselbe Anzahl von Möglichkeiten besteht für das Herausgreifen von $(m-k)$ weißen und $(m+k)$ schwarzen Körnern.)

Als Maß der Wahrscheinlichkeit, welche dafür besteht, daß die Anzahl der herausgegriffenen weißen Körner um höchstens p vom m abweicht, erhält man demnach den Ausdruck

$$N_0 + 2N_1 + \ldots + 2N_p .$$

Setzt man hier überall für n den Wert $h \cdot m$ ein, so sieht man, daß der Ausdruck (bei gegebenem h) nur vom m und p abhängt. Zur Bestimmung der Wahrscheinlichkeit, daß ***nahezu*** m weiße und m schwarze Körner herausgenommen sind, haben wir nun für p einen Wert zu setzen, der klein ist gegen m. Wir treffen die Normierung, indem wir für p die größte ganze Zahl unterhalb $m^{3/5}$ nehmen, so daß p/m kleiner ist als

$$\frac{1}{m^{2/5}}.$$

Der Wert der Wahrscheinlichkeit hängt dann nur noch von m ab, und es läßt sich durch Ausführung einer Abschätzungsrechnung beweisen, daß er dem Grenzwert 1 zustrebt, wenn m ins Unendliche wächst. Es ist demnach die Wahrscheinlichkeit, mindestens $(m - m^{3/5})$ und höchstens $(m + m^{3/5})$ weiße Körner zu greifen, für eine genügend große Gesamtzahl der herausgenommenen Körner beinahe gleich 1; und man kann daher sagen: es ist bei einer sehr großen Zahl von Körnern, so wie wir sie in den praktischen Fällen immer haben, fast sicher, daß man beim Herausgreifen ungefähr ebensoviele weiße wie schwarze Körner bekommt.

In dieser Weise findet jenes einfache Experiment seine Erklärung. Es handelt sich hier nicht um ein eigentliches Naturgesetz, sondern um ein Wahrscheinlichkeitsgesetz. Die mathematische Ableitung für sich allein führt noch nicht zur Voraussage des zu erklärenden Effektes, sondern es muß noch das Erklärungsprinzip hinzugenommen werden, daß dasjenige, was in fast allen Fällen eintritt, auch tatsächlich beobachtet wird. Das Ergebnis einer solchen Schlußfolgerung kann daher auch nicht die einem wahren Naturgesetze zukommende ausnahmlose Gültigkeit beanspruchen, vielmehr gilt es nur in dem Sinne wie etwa der Satz, daß beim Roulettespiel nicht 100 mal hintereinander rot fällt.

Wenngleich somit die durch Wahrscheinlichkeit begründeten Sätze die Möglichkeit einer Ausnahme nicht grundsätzlich ausschließen, so ist doch die Methode der Wahrscheinlichkeitstheorie für die Erklärung der Naturerscheinungen nicht minder bedeutungsvoll als die Methode der Ableitung aus den eigentlichen Naturgesetzen, welche für sich noch nicht ausreicht, um uns die Erfahrungstatsachen verständlich zu machen. Um nur ein Beispiel zu nennen, so kann die Tatsache, daß zwei gleich starke Kerzen doppelt so stark leuchten wie eine einzelne Kerze, durch die Gesetze der Optik nicht erklärt werden, sondern es bedarf dazu einer *statistischen* Überlegung.

Die Methode der Wahrscheinlichkeitsbetrachtung hat eine große Verbreitung in der Physik erlangt. Ihr wichtigstes Anwendungsgebiet bildet die kinetische Gastheorie. Hier besteht im Vergleich zu dem Sandmischungsproblem insofern eine wesentlich größere Schwierigkeit, als man es nicht mit reinen Massenerscheinungen zu tun hat, sondern die Gesetze der Mechanik mit hineinspielen. Man denkt sich ein Gas als bestehend aus sehr vielen, ungefähr gleichen Molekülen, die etwa als elastische Kugeln vorzustellen sind, welche nach mannigfachen Richtungen hin und her fliegen und dabei Zusammenstöße erfahren. Die einzelnen Kugeln bewegen sich nach den Gesetzen der Mechanik, wobei die Gravitationswirkungen vernachlässigt werden können;

d. h. bei freier Bahn ist die Bewegung geradling gleichförmig und beim Zusammenprall zweier Kugeln kommen die Stoßgesetze zur Anwendung. Für die Ableitung der Gasgesetze brauchen wir uns nun nicht darum zu kümmern, wo die einzelnen Moleküle bleiben, sondern es interessieren uns nur gewisse Mittelwerte. Bei der Bestimmung dieser Mittelwerte müssen aber die Gesetze der Bewegung mit berücksichtigt werden. Es kommen daher in der Gastheorie Differentialgleichungen und Zufallsgesetze gemeinsam zur Geltung.

Die enormen mathematischen Schwierigkeiten, welche hieraus entspringen, werden vielfach nicht beachtet, und es ist eine verbreitete Meinung, daß die heute noch bestehende Unvollkommenheit in der Begründung der kinetischen Gastheorie ganz auf philosophischem Gebiet liege, nämlich die grundsätzliche Zulässigkeit der wahrscheinlichkeitstheoretischen Deutung physikalischer Gesetze betreffe. Dieses Übersehen der mathematischen Schwierigkeiten rührt inbesondere davon her, daß man den Begriff des *Wahrscheinlichen* oft nicht genügend auseinanderhält von dem des *Vermutlichen*. Ich will daher auf diesen Punkt etwas näher eingehen.

In der Mathematik ist der Begriff des Vermutlichen ohne weiteres klar. Es gibt hier manche Sätze, die man zwar nicht beweisen kann, von denen man aber vermutet, daß sie richtig sind und daß es nur einer geeigneten Gedankenkombination bedarf, um den Beweis für sie zu erbringen. Von solchen Sätzen kann man sagen, daß sie vermutlich richtig sind. Zum Beispiel gilt vermutlich der Satz, daß die Gleichung

$$p - q = 2$$

für unendlich viele Primzahlpaare p, q erfüllt ist. Hiermit soll gesagt sein, daß die Erfahrungen, welche man bei der Beschäftigung mit den Zahlen gemacht hat, die Annahme jenes Satzes nahelegen. In entsprechendem Sinne läßt sich behaupten, daß vermutlich $2^{\sqrt{2}}$ eine irrationale Zahl ist. Ein anderes berühmtes Beispiel ist die schon früher von mir erwähnte Riemannsche Vermutung über die Verteilung der Nullstellen der Funktion $\zeta(z)$.

Bei all diesen Fällen ist von Wahrscheinlichkeit keine Rede. Es wäre gänzlich verfehlt, wenn man Sätze wie die eben genannten durch Wahrscheinlichkeitsbetrachtungen plausibel machen wollte in dem Sinne, daß man etwa zeigte, es gebe mehr Möglichkeiten für die Gültigkeit als für die Ungültigkeit des betreffenden Satzes. Aus solchen Gründen könnte z. B. das Zutreffen der Riemannschen Vermutung gewiß nicht als wahrscheinlich, sondern höchstens als sehr unwahrscheinlich erklärt werden.

Nun spielen auch in der Physik die Vermutungen eine große Rolle. Man denke hier nicht etwa an die physikalischen Hypothesen, welche versuchsweise der Erklärung von Tatsachen zugrunde gelegt werden, wie z. B. Annahmen über die Struktur des Wasserstoffatoms. Sondern es handelt sich um Vermutungen im eigentlichen Sinne, und zwar Vermutungen mathematischen Charakters. Ein einfacher Fall der Anwendung einer solchen Vermutung liegt vor, wenn der Physiker bei der Lösung einer Differentialgleichung, deren genaue Integration zu schwierig oder zu umständlich wäre, eine Vereinfachung vornimmt durch Weglassen von Gliedern, deren Betrag bei dem zu untersuchenden Vorgang so gering ist, daß sie auf das Endresultat keinen merklichen Einfluß haben. Hier müßte eigentlich die Zulässigkeit der Vernachlässigung durch eine genaue Abschätzung nachgewiesen werden. Das geschieht aber für gewöhnlich nicht, sondern man begnügt sich mit einer gefühlsmäßigen Überzeugung. Bei diesen und ähnlichen Vernachlässigungen pflegt es sich so zu verhalten, daß eine strenge Durchführung der Abschätzung wohl praktisch durchführbar wäre, aber für überflüssig befunden wird.

Anders liegt es in der Astronomie bei den Rechnungen der *Störungstheorie.* Hier wendet man vielfach semikonvergente Reihen an, wobei man so viele Glieder für die Rechnung berücksichtigt, daß das nächste Glied auf die letzte zu berechnende Dezimalstelle keinen Einfluß mehr hat. Wollte man hier streng verfahren, so würden die Rechnungen so kompliziert, daß man sie nicht ausführen könnte. Das geschieht also nicht, sondern man verläßt sich darauf, daß die berücksichtigten Glieder das der Theorie entsprechende Ergebnis (innerhalb der vorgeschriebenen Fehlergrenzen) liefern, und ist zufrieden, wenn die Beobachtung den berechneten Werten entspricht. In diesen Fällen ist also die Bestätigung der Newtonschen Theorie nur eine vermutliche. Noch weitergehend ist die Anwendung von Vermutungen in der *kinetischen Gastheorie.* Hier werden Sätze benutzt, die sich überhaupt nicht durch Rechnung bestätigen lassen, und von deren Beweis man weit entfernt ist.

Den Unterschied in der Schwierigkeit der Aufgaben, welche bei den drei genannten Arten von Vermutungen vorliegen, kann ich wohl am besten durch den Vergleich mit drei mathematischen Aufgaben erklären. Ich wähle folgende Beispiele:

1. die Berechnung der 10^5-ten Dezimalstelle von π (entsprechend dem Fall der abgekürzten Integration). Hier würde die Rechnung zwar sehr mühsam, aber doch ausführbar sein.

2. die Berechnung der $10^{(10^{10})}$-ten Dezimalstelle von π (entsprechend dem Fall bei der Störungsrechnung). Hier ist an eine wirkliche Ausführung der Rechnung nicht zu denken; wohl aber läßt sich für die Zahl der erforderlichen Rechenoperationen eine obere Schranke angeben.

3. der Beweis der Limesgleichung

$$\lim_{n=\infty} \frac{A_n}{n} = \frac{1}{10},$$

worin A_n die Anzahl der Ziffern 1 in den ersten n Stellen der Dezimalbruchentwicklung von π bedeutet, d. h. also der Beweis des Satzes, daß in der Dezimalbruchentwicklung von π die relative Häufigkeit des Auftretens der Ziffer

1, asymptotisch gerechnet, 1/10 beträgt (entsprechend dem Fall der Anwendung mathematischer Theoreme in der Gastheorie). Hier könnte selbst durch beliebig viele Rechner der Satz nicht nachgeprüft werden. Man ist auch bisher nicht imstande, den Satz zu beweisen, man vermutet nur, daß er richtig ist.

Auf solchen nur vermutlich richtigen Sätzen von der Art wie der eben angeführte beruht das ganze Gebäude der kinetischen Gastheorie. Über diesen Umstand täuscht man sich deshalb oft hinweg, weil die betreffenden Sätze der Wahrscheinlichkeitstheorie angehören und infolgedessen die bloß wahrscheinliche (d. h. vermutliche) Gültigkeit der Sätze leicht verwechselt wird mit derjenigen Wahrscheinlichkeit, von der die Sätze handeln. Es soll hiermit das Verfahren der kinetischen Gastheorie keineswegs angefochten werden. Vielmehr habe ich ja bereits bei der Erörterung der Methode der Mathematik betont, daß es ein durchaus berechtigtes Verfahren ist, wenn man beim Aufbau einer speziellen Theorie gewisse noch unbewiesene, aber plausible Sätze (als Axiome) voranstellt, sofern man sich nur über die Unvollständigkeit der Begründungsweise im klaren ist.

In welcher Weise die Vermutung von Sätzen über Wahrscheinlichkeiten in der Gastheorie eine Rolle spielt, will ich des näheren an einem Beispiel ausführen. Es sei ein Kasten durch eine Wand in zwei Fächer A, B geteilt. Das Fach B sei anfangs leer, während A mit einem Gase erfüllt sei. Nun werde die trennende Wand weggenommen. Dann erfüllt nach Ablauf einer gewissen Zeit das Gas die beiden Teile A, B des Kastens mit gleicher Dichtigkeit. Dies ist eine Erfahrungstatsache, und es soll nun vom Standpunkt der kinetischen Gastheorie eine Erklärung dafür gegeben werden. Hierzu müssen wir ausgehen davon, daß das Gas aus einer Anzahl n von gleichbeschaffenen Molekülen besteht. Zur Anfangszeit t_1 befinden sich alle Moleküle in A; jedes hat eine bestimmte Lage und Geschwindigkeit.

$$u_1, v_1, w_1, \ldots, u_n, v_n, w_n,$$

seien die Komponenten der Anfangsgeschwindigkeit für die n Moleküle, so daß, wenn E die Energie des Gases (d. h. die gesamte kinetische Energie der Moleküle) und m die Masse eines Moleküls bedeutet, die Gleichung

$$\sum_{k=1}^{n}(u_k^2 + v_k^2 + w_k^2) = \frac{2E}{m}$$

besteht. Deutet man die $3n$ Geschwindigkeitskomponenten als Koordinaten eines n-dimensionalen (euklidischen) Raumes, so wird jede Verteilung der Anfangsgeschwindigkeiten durch einen Punkt dieses Raumes dargestellt, und die Gesamtheit der möglichen Verteilungen, welche einem festen Wert der Energie E entspricht, stellt sich dar als die Gesamtheit der Punkte auf einer $(3n-1)$-dimensionalen Kugelfläche.

Denken wir uns den Versuch bei Zimmertemperatur mit einem Kasten von etwa 1000 ccm Ausdehnung und mit gewöhnlicher Luft ausgeführt, so bestimmt sich die Größenordnung der Werte von n, m und E gemäß anderweitig (durch Vergleichung von Theorie und Experiment) gewonnenen Ergebnissen der kinetischen Gastheorie. Zur Anfangszeit t_1 werde die trennende Wand beseitigt. Durch die Anfangslagen und Anfangsgeschwindigkeiten der Moleküle ist dann der Verlauf der Zustandsänderungen völlig bestimmt, und es läßt sich (wenigstens grundsätzlich) berechnen, wie groß zu einer gegebenen (hinter t_1 liegenden) Beobachtungszeit t_2 die Anzahl n_A der Moleküle in A und die Zahl n_B derjenigen in B ist.

Der Einfachheit halber wollen wir annehmen, daß die beiden Fächer A und B von gleicher Größe sind. Die Bedingung dafür, daß zur Zeit t_2 gleiche Dichte des Gases in A und B beobachtet wird, besteht dann darin, daß die Abweichung des Verhältnisses $u_A : n$ von dem Werte $1/2$ eine gewisse Grenze g nicht übersteigt. Um nun die Tatsache, daß man immer nach einer gewissen Zeit gleiche Dichte in A und B vorfindet, vom Standpunkt der statistischen Theorie zu erklären, würde es ausreichen, zu beweisen, daß für gegebene Werte von n, m, E (von der richtigen Größenordnung), bei beliebig in A vorgeschriebenen Anfangslagen der Moleküle und für einen beliebig, nicht zu kurz, gewählten Zeitabstand $(t_2 - t_1)$ die Bedingung der gleichen Dichte

$$\left| \frac{n_A}{n} - \frac{1}{2} \right| \leq g$$

fast immer, d. h. bei fast allen Verteilungen der Anfangsgeschwindigkeiten

$$u_1, v_1, w_1, \ldots, u_n, v_n, w_n,$$

für welche

$$\left| \frac{n_A}{n} - \frac{1}{2} \right| > g$$

ist, bei der Darstellung auf der $(3n-1)$-dimensionalen Kugelfläche nur einen ganz kleinen Flächenteil einnehmen.

Dieser rein mathematische Satz würde uns die gewünschte Erklärung liefern im Sinne des Prinzips, daß dasjenige, was fast immer eintrifft, wirklich beobachtet wird. Bisher hat aber niemand den Satz bewiesen, und das ist auch äußerst schwierig. Trotzdem haben wir guten Grund, den Satz als richtig zu vermuten. Wir können nämlich die Bedingungen des Satzes so weit schematisch vereinfachen, daß sich alles vollkommen überblicken läßt.

Eine solche Vereinfachung erhalten wir, wenn wir von der Ausdehnung der Moleküle absehen, also die Moleküle als punktförmig betrachten und außerdem nur eine Raumdimension in Betracht ziehen. Wir haben dann eine Anzahl n von Massenpunkten, die sich auf einer Strecke bewegen, an deren

Enden sie reflektiert werden. A und B sind die beiden Hälften der Strecke. Die Anfangsgeschwindigkeit eines Massenpunktes wird durch je eine Zahl gegeben, und die möglichen Wertsysteme der Anfangsgeschwindigkeit bei gegebener Gesamtenergie stellen sich dar durch die Punkte einer $(n-1)$-dimensionalen Kugelfläche. Hinsichtlich der Anfangslagen genügt es für das Prinzipielle, den Spezialfall zu betrachten, daß sich zu Anfang alle Massenpunkte an dem einen Ende der Strecke befinden.

Die wesentliche Vereinfachung beruht nun darauf, daß die Zusammenstöße der Moleküle keine Schwierigkeit mehr bieten. Das Stoßgesetz besagt nämlich jetzt, daß die Massenpunkte beim Zusammentreffen ihre Geschwindigkeiten vertauschen, und das kommt im Ergebnis auf dasselbe hinaus, wie wenn die Massenpunkte ungehindert durch einander hindurch gingen. Das heißt, wir brauchen die Zusammenstöße überhaupt nicht zu berücksichtigen. Aufgrund dieses Umstandes läßt sich die Rechnung vollkommen durchführen. Nimmt man als Länge der Strecke 10 cm, für n die Zahl 10^{20} (die Anzahl der Moleküle im Kubikcentimeter bei einem wirklichen Gas ist etwa $27 \cdot 10^{18}$), und für $2E/m.$, d. h. für die Summe der Quadrate der Anfangsgeschwindigkeiten, in cm/sec. gerechnet, den Wert

$$(5 \cdot 10^4)^2 n = 25 \cdot 10^{28},$$

so ergibt sich, daß nach Ablauf von 6 Minuten gleiche Dichte in beiden Hälften der Strecke mit der Genauigkeit

$$\left| \frac{n_A}{n} - \frac{1}{2} \right| < \frac{1}{1000}$$

vorliegt, sofern man von gewissen Wertsystemen der Anfangsgeschwindigkeiten absieht, welche bei der Darstellung auf der $(n-1)$-dimensionalen Kugelfläche einen Bruchteil der Gesamtfläche ausmachen, der kleiner ist als

$$2 \cdot 10^{-6}\, e^{-10^{14}},$$

also ein winzig kleiner Bruchteil. (Die auszunehmenden Wertsysteme der Anfangsgeschwindigkeiten hängen natürlich von der Beobachtungszeit t_2 ab.) Die Genauigkeit

$$\left| \frac{n_A}{n} - \frac{1}{2} \right| < \frac{1}{100}$$

für die Gleichverteilung stellt sich bereits nach 4 Sekunden ein.

Diese Einwendungen betreffen gerade die Frage, um derentwillen wir die Anwendung des Wahrscheinlichkeitsbegriffs im Gebiete der Physik genauer in Betracht gezogen haben, nämlich die Frage der Auszeichnung einer Zeitrichtung. Die Erscheinungen, bei denen sich besonders der Erfolg der statistischen Methode zeigt, haben den Charakter der *Irreversibilität.* So handelt es sich bei dem ausgeführten Beispiel um die Erklärung der Tatsache, daß ein Gas sich von selbst so weit wie möglich ausdehnt, (während der umgekehrte Prozess des Zusammenschrumpfens nicht beobachtet wird.).

Andererseits enthalten die Grundlagen der kinetischen Gastheorie keinerlei Auszeichnung einer Zeitrichtung. Denn es werden ja nur die Gesetze der Mechanik und die Methoden der Wahrscheinlichkeitstheorie benutzt. In diesen beiden treten aber Vergangenheit und Zukunft ganz symmetrisch auf. Die Gesetze der Mechanik bleiben ungeändert, wenn man t durch $-t$ ersetzt,* und es sind daher alle rein mechanischen Vorgänge reversibel; d. h. wenn bei einem mechanischen Bewegungsvorgang in einem Moment alle Geschwindigkeiten ihrer Richtung nach umgekehrt würden, so müßte sich alles rückläufig bewegen. Und in der Idee der statistischen Methode liegt gewiß nichts, was eine Zeitrichtung auszeichnet. Wie soll da die Erklärung eines irreversiblen Vorganges durch die kinetische Gastheorie möglich sein?

Es scheint, daß hier ein Fehlschluß vorliegen muß. Hierauf ist zunächst zu erwidern, daß bei den Erklärungen irreversibler Vorgänge durch die kinetische Theorie die Unsymmetrie in bezug auf Vergangenheit und Zukunft durch die Art der *Anfangs- und Randbedingungen* zustande kommt. Ein Anfangszustand wird als gegeben betrachtet, von welchem an das Gas sich selbst überlassen sein soll, während vor dem Anfangszustand das Gas irgendwie beeinflußt sein kann. Bei dem betrachteten Beispiel ist dies ganz deutlich. Hier wird angenommen, daß bis zu der Anfangszeit t_1 die Gasmoleküle in der Hälfte A des Kastens eingeschlossen sind, von der Zeit t_1 an sich aber in dem ganzen Kasten frei bewegen können. Bei dieser Problemstellung treten von vornherein Vergangenheit und Zukunft in verschiedener Weise auf, und es ist daher kein Wunder, daß in dem Ergebnis der theoretischen Überlegung eine Zeitrichtung ausgezeichnet ist.

Die mechanische und wahrscheinlichkeitstheoretische Betrachtung würde genau so gut auch folgenden Satz liefern: Wenn ein Gas, das sich in einem Kasten befindet, von einer Zeit t an durch eine Scheidewand auf eine Hälfte des Kastens beschränkt wird, während es vorher innerhalb des Kastens unbeschränkt sich selbst überlassen ist, so sind zu jeder Zeit, welche gegen t um einen gewissen Mindestabstand zurückliegt, in fast allen Fällen die beiden Hälften des Kastens mit gleicher Dichte von dem Gase erfüllt. Nur kann man diesen Satz nicht verwerten, weil die hierbei vorausgesetzten Randbedingungen in der Wirklichkeit nicht vorgefunden werden und sich auch nicht künstlich herstellen lassen. (Man kann nicht abwarten, daß das Gas einmal auf die Hälfte des verfügbaren Raumes zusammenschrumpft, um in dem betreffenden Augenblick die Scheidewand einzusetzen.) Es kommt also bei der Erklärungsweise der kinetischen Gastheorie wesentlich auch darauf an, was

* Eine Folgerung aus dem allgemeinen Relativitätsgesetze von Einstein.

für Randbedingungen in der Natur verwirklicht sind oder sich willkürlich hervorbringen lassen.

Durch diesen Gesichtspunkt erledigt sich auch eine zweite Fassung des Einwandes gegen die Gastheorie. Diese besteht in folgender Argumentation: Da jeder Gasprozeß ein mechanischer Vorgang ist, so ist auch der zu ihm entgegengesetzt verlaufende Prozeß physikalisch möglich, und zwar geht er aus dem ersten hervor, indem in einem Augenblick alle *Geschwindigkeiten umgekehrt* werden. Nun kann sich doch aber durch die Umkehrung der Geschwindigkeiten an der Wahrscheinlichkeit nichts ändern; das würde ja der Symmetrie widerstreiten. Durch Wahrscheinlichkeitsgründe kann daher nicht erklärt werden, warum bei einem Gasprozeß eine Richtung des Ablaufes vor der entgegengesetzten etwas voraus hat, warum also z. B. ein Gas sich nicht ebensogut von selbst zusammenziehen kann, wie es sich von selbst ausdehnt.

Bei dieser Überlegung wird wiederum übersehen, daß es für die Beurteilung des wirklichen Vorkommens eines Prozesses (gemäß der Methode der kinetischen Gastheorie) nicht auf die Wahrscheinlichkeitsgrößen allein, sondern auch auf die Realisierbarkeit gewisser Randbedingungen ankommt. Nun wissen wir, daß die zu dem Prozeß der Ausdehnung eines Gases gehörige Anfangsbedingung sich sehr gut realisieren läßt (und auch vielfach von selbst in der Natur verwirklicht ist.) Daraus folgt aber keineswegs, daß auch zu dem umgekehrten Vorgang eine zugehörige Anfangsbedingung tatsächlich herstellbar ist. Ich kann ja den Zustand eines Gases nicht in der Weise beeinflussen, daß ich in einem Augenblick jedem einzelnen Molekül einen bestimmten Anstoß gebe, so daß seine Geschwindigkeit sich umkehrt. Dazu müßte ich eine Schar von mikroskopischen Wesen (Maxwellsche Dämonen) zu meiner Verfügung haben.

Es kommen also hier *anthropologische Gründe* mit ins Spiel. Im ganzen ergibt sich aus unserer letzten Erörterung, daß das Problem der Irreversibilität in der Gastheorie hinauskommt auf die Frage der *Realisierbarkeit von Anfangs- oder Randbedingungen* für das mechanische System der Gasmoleküle.

Nun findet sich die Erscheinung der Irreversibilität nicht nur bei den Gasprozessen, sondern auch bei vielen anderen Gattungen physikalischer Vorgänge. Im Gebiete der Elastizität gibt es viele einseitig verlaufende Prozesse, insbesondere die Ausbreitung des Schalles. (Wir können niemals eine Explosion im voraus hören.) Ferner bei der Strahlung ist die Ausbreitung einer Kugelwelle ein typisches Beispiel eines irreversiblen Vorganges. Und die neuerdings entdeckten Prozesse des radioaktiven Zerfalles von Substanzen haben gleichfalls irreversiblen Charakter. Wir werden also veranlaßt, die Frage der Irreversibilität allgemeiner zu untersuchen. Hierzu knüpfen wir an unsere Betrachtungen über die physikalische Gesetzlichkeit an. Wir stellten da fest, daß der Ablauf des physikalischen Geschehens sich gemäß gewissen für die Zustandsgrößen geltenden Differentialgleichungen vollzieht, daß diese aber zur Bestimmung des Geschehens nicht ausreichen, da die Lösungen der Differentialgleichungen noch willkürliche Konstanten, bzw. willkürliche Funktionen enthalten.

Wenn es möglich wäre, diese Unbestimmtheit durch ein System von universalen (das Weltganze betreffenden) Zusatzbedingungen zu beheben, so würden wir, wie schon erwähnt, jene Hegelsche Physik erhalten, durch welche grundsätzlich alles in der Natur theoretisch zu bestimmen sein müßte. Lassen wir diesen phantastischen, und auch aus philosophischen Gründen anfechtbaren Gedanken fallen, so bleibt nichts anderes übrig, als das Weltgeschehen in vierdimensionale Stücke zu zerlegen. Statt der Welt als Ganzes betrachten wir nur einzelne Stücke, wo die zusätzlichen Randbedingungen sich in einfacher Weise angeben lassen. Von diesem Standpunkt haben wir hinsichtlich der Möglichkeit eines Prozesses zweierlei zu unterscheiden, nämlich die Vereinbarkeit mit den Differentialgleichungen und die Realisierbarkeit der Randbedingungen, auf welcher die Existenzmöglichkeit für die einzelne Lösung der Differentialgleichungen beruht.

Prüfen wir nun nach dieser vorbereitenden Überlegung die Frage, ob in dem physikalischen Geschehen eine Zeitrichtung ausgezeichnet ist, so müssen wir zunächst feststellen, daß bei den Differentialgleichungen jedenfalls keine solche Auszeichnung vorliegt. Von den Gleichungen der Mechanik habe ich bereits hervorgehoben, daß sie bei der Vertauschung von t mit $-t$ ungeändert bleiben. Aber auch in der Feldtheorie, wo die partiellen Differentialgleichungen das Fundament bilden, besteht eine solche Symmetrie: Die Maxwellschen Grundgleichungen der Elektrodynamik ändern sich nicht, wenn darin x, y, z, t durch $-x, -y, -z, -t$ ersetzt werden. Die Grundgleichungen der Physik haben also die Eigenschaft, daß man aus jeder ihrer Lösungen durch Umkehrung der Zeitrichtung (und eventuell noch Änderung des räumlichen Windungssinnes) wieder eine Lösung erhält. Aber natürlich ist das *nicht dieselbe* Lösung. Und wenn jene sich realisieren läßt, so braucht das von dieser nicht zu gelten.

Wenn wir nun zusehen, wie es tatsächlich hiermit steht, so finden wir allerdings viele Lösungen verwirklicht, von welchen die zeitlich umgekehrten Lösungen in Wirklichkeit nicht vorkommen. Aber diese Fälle haben immer nur den Charakter des physikalisch Zufälligen. Es sprechen dabei vor allem anthropologische Gründe mit. Besonders deutlich tritt dies bei der *Mechanik* hervor. Hier wäre grundsätzlich jede Lösung realisierbar. Nur ist die willkürliche Herstellung von Anfangsbedingungen durch unsere menschliche Veranlagung beschränkt. Wir können im allgemeinen nur solchen Körpern eine willkürliche Lage und Geschwindigkeit erteilen, deren Volumen von derselben Größenordnung ist wie der menschliche Leib. Dagegen die

Gestirne können wir nicht aus ihrer Bahn bringen, und ebensowenig können wir die Moleküle eines Gases einzeln dirigieren. Und so finden wir einerseits bei den astronomischen Vorgängen, andererseits bei den Gasprozessen die Zeitrichtung nach der Zukunft wesentlich unterschieden von der Richtung nach der Vergangenheit. Die Irreversibität, welche man hier vorfindet, ist also sozusagen nur eine Scheinbare.

Bei den Gasen läßt sich dies sogar direkt zeigen; wir finden nämlich bei einem stationären Gas die Gleichmäßigkeit der Dichteverteilung nur in dem Grade vor, wie es (aufgrund der Molekularhypothese) durch die Gesetze der Wahrscheinlichkeit verlangt wird. Tatsächlich ist die Verteilung der Dichte in einem Gas mit Schwankungen behaftet, welche sich ihrer Größe nach durch die kinetische Gastheorie bestimmen lassen. Wir treffen demnach bei der Beobachtung von hinreichend kleinen Raumstücken in einem Gas auch solche Veränderungen an, welche zu den im Großen stattfindenden entgegengesetzt sind. Es liegt also bei dem Zustandekommen der gleichmäßigen Dichte eines Gases keine strenge Irreversibilität vor, sondern nur eine uneigentliche, welche auf der leichteren Realisierbarkeit gewisser Anfangsbedingungen beruht. (Ich will hier noch erwähnen, das durch die Schwankungen in der Dichte der Atmosphäre die blaue Farbe des Himmels sich erklären läßt. Daß diese Erklärung sachgemäß ist, zeigt sich insbesondere daran, daß die Beziehung zwischen der Intensität des Himmelsblau und der mittleren Dichte der Atmosphäre, welche sich dabei als Folgerung ergibt, mit den Beobachtungen im Einklang steht.)

Ähnlich wie mit den Gasprozessen verhält es sich mit vielen anderen Vorgängen, welche dem Bereich der Mechanik angehören. So stellt sich z. B. die in einer Wasserschüssel durch Hineinwerfen eines Steines erzeugte Wellenausbreitung als ein irreversibler Vorgang dar. Prinzipiell muß es aber auch möglich sein, die entgegengesetze Wellenbewegung vom Rande der Schüssel her so zu erzeugen, daß ein im Wasser befindlicher Stein dadurch aus dem Wasser herausgeschleudert wird. Es ist nur eine zufällige Unfähigkeit, wenn wir nicht imstande sind, eine so kunstvolle Einrichtung zu treffen, um den erforderlichen Schwingungszustand im Wasser hervorzubringen.

Im Gebiete der *elektromagnetischen Feldtheorie* können wir zwar nicht ohne weiteres behaupten, daß jedes Integral der Feldgleichung realisierbar sein muß. Aber auch hier finden wir kein Anzeichen einer prinzipiellen Irreversibilität. Allerdings gibt es auch Lösungen der elektrodynamischen Feldgleichungen, von welchen sich die zeitliche Umkehrung nicht realisieren läßt. Ein Beispiel dafür bildet die von einer punktförmigen Lichtquelle ausgehende Strahlung. Aber man sieht hier auch, daß die Realisierung des umgekehrten Vorganges nur darum nicht gelingt, weil die entsprechenden Anfangsbedingungen zu kompliziert sind.

Andererseits haben wir in der Elektrodynamik eine Fülle von Experimenten, bei denen die Reversibilität unmittelbar nachzuweisen ist: Bei einer

Dynamomaschine läuft der Strom im einen oder anderen Sinne je nach der Richtung, in der man den Grammeschen Ring dreht. Die Erwärmung einer Lötstelle von Wismut und Antimon erzeugt einen elektrischen Strom, die Abkühlung einen Strom in entgegengesetzter Richtung. Und so gibt es noch viele Beispiele.

Nun wird durch die Mechanik und die Feldtheorie des Elektromagnetismus nicht die gesamte Physik beherrscht. Insbesondere ist die moderne *Quantentheorie* zu nennen, deren Gesetze sich nicht in den Rahmen jener Theorien einfügen lassen. Und zwar ist dabei die Art der Abweichung von der Mechanik gerade für unsere Betrachtung von Wichtigkeit. Es handelt sich dabei um die Bewegungen der Elektronen im Atom. Nach der Auffassung der Quantentheorie bewegen sich die Elektronen aufgrund der Coulombschen Anziehungs- und Abstoßungskräfte entsprechend den Bewegungsgleichungen der Mechanik. Aber es muß überdies noch eine Zusatzbedingung erfüllt sein, welche darin besteht, daß einige von den Integrationskonstanten der Differentialgleichungen nur gewisse diskrete Werte annehmen können, welche durch ein zahlentheoretisches Gesetz bestimmt sind. Hier tritt also in der Tat zu den Differentialgleichungen noch eine grundsätzlich beschränkende Bedingung für die Lösungen. Aber eine Irreversibilität kommt dadurch nicht zustande; denn in den quantentheoretischen Bedingungen für die Integrationskonstanten ist nichts enthalten, was sich auf eine Zeitrichtung bezieht.

Es bliebe nun noch die Frage, wie es sich bei den *radioaktiven Vorgängen* in betreff der Umkehrbarkeit verhält. Eine Entscheidung hierüber ist gegenwärtig nicht möglich, da die Theorie der Radioaktivität sich erst noch in den Anfängen befindet. Immerhin besteht die Möglichkeit, daß auch bei dem radioaktiven Zerfall die Irreversibilität nur eine Scheinbare ist.

Zusammenfassend können wir sagen, daß in dem physikalischen Geschehen die Auszeichnung einer Zeitrichtung nicht nachweisbar ist. Es gibt zwar irreversible Prozesse; aber beim näheren Zusehen bemerkt man, daß es sich in den betreffenden Fällen nur um eine scheinbare Irreversibilität handelt, welche auf den besonderen Bedingungen des Experimentierens, also auf anthropologischen Gründen beruht. Anders verhält es sich freilich, sobald wir die *organische Natur* in Betracht ziehen. Hier ist die Zeitrichtung von der Geburt bis zum Tode merklich ausgezeichnet.

Wenden wir uns nun zu unserer Erörterung über die *Kausalität* zurück, von welcher wir ja auf die Frage der Auszeichnung einer Zeitrichtung gekommen waren. Wir hatten uns die Frage vorgelegt, ob nicht bei der physikalischen Kausalität zu dem, was durch den Charakter der Differentialgleichungen zum Ausdruck gebracht wird, noch etwas hinzukomme, insbesondere ob nicht das Zeitverhältnis bei der Kausalität, in dem Sinne, daß die Ursache das Frühere, die Wirkung das Spätere ist, noch eigens berücksichtigt werden müsse. Man könnte denken, daß diese Frage bereits durch die Relativitätstheorie im verneinenden Sinne entschieden werde, weil es ja nach dieser gar keine eindeutige zeitliche Reihenfolge der Ereignisse gibt. Dies wäre jedoch ein Irrtum. In der Tat wird die Möglichkeit einer für das Kausalverhältnis wesentlichen Zeitrichtung von der Relativitätstheorie nicht angefochten. Denn diese behauptet die Unbestimmtheit des Zeitverhältnisses nur für solche Ereignisse, die in keinerlei Wirkungszusammenhang miteinander stehen können.

Ziehen wir nun aber das Ergebnis unserer Betrachtungen über die Irreversibiltät heran, so gelangen wir zu der Auffassung, daß für die Erklärung des physikalischen Geschehens die Bestimmung des Zukünftigen durch das Vergangene prinzipiell nichts voraus hat von der Bestimmung des Vergangenen durch das Zukünftige. Nur für uns Menschen hat die theoretische Ableitung des Zukünftigen aus dem Vergangenen eine weit größere Wichtigkeit, weil wir über die Vergangenheit durch Erinnerung und Mitteilung wissen, während uns die Zukunft unbekannt ist, und weil andererseits unser willkürliches Handeln nur die Zukunft modifizieren kann. Es bildet für uns in der Tat die Vergangenheit des Gegebene, die Zukunft das zu Bestimmende. Das physikalische Geschehen für sich ist aber von diesem Unterschiede unberührt. Überhaupt bedeutet die Einführung des ursächlichen Momentes in die Physik nur eine spezielle Art der Darstellung und Gruppierung der Zusammenhänge.

Der Anlaß zu der ursächlichen Betrachtung in der Physik liegt darin, daß bei den speziellen zu behandelnden Problemen sich zumeist in dem Ausdruck der Differentialgesetze oder der Randbedingungen verschiedene Bestandteile voneinander abheben, von denen für gewöhnlich nur einige, je nach den obwaltenden Größenverhältnissen, zur Geltung kommen. Wenn wir ein einzelnes Ereignis oder die Anwesenheit eines Gegenstandes (bzw. Umstandes) als Ursache einer Veränderung ansehen, so kommt das darauf hinaus, daß wir einen von jenen sich abhebenden Bestandteilen eines Differentialgesetzes oder einer Randbedingung abgesondert betrachten und den Beitrag untersuchen, den er zu dem Gesamtergebnis liefert.

So betrachten wir z. B. bei der Bewegung der Erde um die Sonne den Jupiter als Ursache von Störungen der elliptischen Bahn, indem wir uns zunächst den Jupiter wegdenken und dann erst nachträglich zusehen, welche Änderungen sich ergeben, wenn die auf den Jupiter bezüglichen Glieder (und die hinzutretenden Gleichungen) in dem Kraftgesetz hinzugenommen werden.

Beim Fallen eines Körpers auf die Erde unterscheiden wir als Ursachen die Schwerkraft, die durch die Drehung der Erde bedingte Zentrifugalkraft und den Luftwiderstand. Die vollständige mechanische Theorie des Vorgangs würde alle diese Umstände auf einmal berücksichtigen, indem sie die Bewegung von vornherein auf ein Inertialsystem beziehen und auf die molekulare Struktur des Luftmediums eingehen müßte. Für die Zwecke der Forschung und der Praxis ist es aber sehr nützlich, jene drei Momente aus dem physikalischen Zusammenhang einzeln herauszuheben; denn die Drehung der Erde um ihre Achse macht sich bei dem Vorgang des Falles meist nur in geringen Grade geltend, und die molekularen Vorgänge in der Luft brauchen wir nur ganz summarisch in Anschlag zu bringen; dagegen das Gesetz des freien Falles müssen wir genau kennen, weil es uns den Typus der Bewegung anzeigt und näherungsweise auch deren quantitativen Verlauf.

Die Einführung des Begriffes der Ursache in die Physik geschieht also aus Gründen der Zweckmäßigkeit. Es kommt aber dabei nicht etwas grundsätzlich Neues zu der durch die Differentialgleichungen (und eventuell noch durch Randbedingungen) charakterisierten physikalischen Gesetzlichkeit hinzu. Der Sachverhalt ist hier ähnlich wie bei der Anwendung des *Kraftbegriffes.* Wenn wir von Kräften in der Physik sprechen, so ist das nur eine besondere Ausdrucksform für die Differentialgesetze.

Besonders deutlich zeigt sich dies in der Elektrodynamik. Hier bilden die Maxwellschen Gleichungen zusammen mit dem Lorentzschen Kraftgesetz die Differentialgesetze. Man kann nun die Maxwellschen Gleichungen (bei geeigneten Randbedingungen für das Unendliche) integrieren und die Integrale in das Kraftgesetz einführen. Das Ergebnis, welches man so erhält, läßt sich in dem Sinne deuten, daß zwischen den einzelnen Elektronen eine (mit Lichtgeschwindigkeit sich ausbreitende) Fernkraft wirkt. Man gelangt also zu einem Elementargesetz der elektrischen Wirkung, welches allerdings außerordentlich kompliziert ist. (Die Größe der Kraft eines elektrischen Teilchens auf ein anderes hängt danach von der Bewegungsrichtung und der Beschleunigung der beiden Teilchen ab.) Dieses von Schwarzschild berechnete Kraftgesetz genügt für sich allein, um darauf die Elektrodynamik aufzubauen, wobei man also die Feldgleichungen beiseite lassen kann. Bei diesem Verfahren würde der Feldbegriff ganz eliminiert und alles würde auf die zwischen den elektrischen Teilchen wirkenden Kräfte zurückgeführt. Hierin würde aber nichts mehr liegen als eine für manche Zwecke geeignete Auffassungsweise der elektrodynamischen Gesetze, welche in einfacherer Weise durch die Gleichungen der Feldtheorie formuliert werden. (Das Gesetz der Fernwirkung zwischen den elektrischen Teilchen hat übrigens den grundsätzlichen Mangel, daß es sich nicht unserem Schema der physikalischen Gesetzlichkeit fügt, insofern sich dabei die momentane Änderung des zu einer Zeit bestehenden Zustandes nicht durch diesen Zustand allein eindeutig bestimmt, sondern erst unter Hinzunahme der vorausgehende Zustände.)

Wir kommen somit zu der Ansicht, daß das Ursächliche im engeren Sinne für die Physik keine objektive Bedeutung besitzt, daß vielmehr bei der Aufsuchung der Ursachen die Rücksicht auf die besonderen Bedingungen der menschlichen Wahrnehmung und auf die menschlichen Zwecke wesentlich mitspricht. Es bleibt aber noch die Möglichkeit, daß im Gebiete der belebten Natur das Verhältnis von Ursache und Wirkung die grundsätzliche Bedeutung gewinnt, die man ihm in der Denkweise des täglichen Lebens zuzuschreiben pflegt.

Wenn wir nun aufgrund der von mir dargelegten Auffassung sagen müssen, daß die Betrachtung der physikalischen Vorgänge nach dem Verhältnis von Ursache und Wirkung etwas Anthropomorphes enthält, so ist doch natürlich diese Art des Anthropomorphen, welche nur die Auswahl und die Auffassungsweise der Tatsachen betrifft, wesentlich verschieden von dem eigentlichen Anthropomorphismus, bei welchem der physikalischen Gesetzlichkeit seelische Eigenschaften beigelegt werden, wie es insbesondere da geschieht, wo man in den physikalischen Gesetzen eine *Zielstrebigkeit* der Natur erblickt.

Solche Annahmen von Naturzwecken in der Physik werden insbesondere dadurch veranlaßt, daß sich die Differentialgesetze der Physik zumeist in der Form von Minimalprinzipien aussprechen lassen. So hat man früher die Integralprinzipien der Mechanik, z. B. das Prinzip der kleinsten Wirkung im Sinne einer Zweckverfolgung der Natur gedeutet, und in derselben Weise ist auch das Minimalprinzip der geometrischen Optik aufgefaßt worden, welches besagt, daß die Bahn eines Lichtstrahls in einem Medium von veränderlicher optischer Dichte so verläuft, daß bei Festhaltung der Endpunkte die (durch die örtlich veränderliche Fortpflanzungsgeschwindigkeit des Lichtes bestimmte) Dauer des Lichtweges möglichst gering ist.

Bei diesen Fällen kann man sich allerdings wundern, daß die physikalischen Gesetze durch so einfache Variationsprinzipien ausdrückbar sind. Aber zur Annahme von Naturzwecken liegt hier doch kein Grund vor; eine solche Annahme würde ja auch gar nichts zum Verständnis der besonderen Form der Minimalprinzipien beitragen, während aus den Differentialgesetzen die Gültigkeit der Variationsprinzipien, und zwar in der besonderen ihnen zukommenden Form, sich als notwendige Folgerung ergibt. (Auch muß uns die Bedeutung der Minimalprinzipien durch Einführung von Naturzwecken noch besonders deshalb verdächtig sein, weil ein und dasselbe Differentialgesetz durch verschiedene Minimalforderungen zum Ausdruck gebracht werden kann.)

Als ein Beispiel eines Gesetzes, bei welchem man zu einer anthropomorphen Deutung besonders stark versucht sein könnte, ist die *Bohrsche Quantenregel* bemerkenswert. Wie ich schon erwähnte, nimmt man in der Quantentheorie an, daß die Elektronen im Atom sich nur auf speziellen Bahnen um den Kern bewegen können, welche dadurch bestimmt sind, daß einige von den Integrationskonstanten der Bewegungsgleichungen gewisse zahlentheoretische Bedingungen erfüllen. Im einfachsten Falle hat man eine einparametrige Schar von Kreisbahnen, und wenn n den ganzzahligen Parameter der Schar bedeutet, so ergibt die Berechnung der Energie für die n-te Bahn einen Ausdruck von der Form C/n^2 (wo C von n unabhängig ist).

Nun behauptet die Quantentheorie ferner, daß ein Elektron von einer Umlaufsbahn auf eine andere überspringen kann und daß hierbei verlorene oder gewonnene Energie in Form einer monochromatischen Strahlung abgegeben, bzw. aufgenommen wird. Dabei bestimmt sich die Frequenz ν der Strahlung aus der Bohrschen Regel, nach welcher die Differenz zwischen der Anfangs- und der Endenergie des Elektrons gleich dem Produkt $h \cdot \nu$ ist, wo h eine universelle Konstante (das Plancksche Wirkungsquantum) bedeutet. Bei dem genannten einfachen Fall der Kreisbahnen ergibt die Regel für den Sprung von der n-ten Bahn auf die m-te die Gleichung

$$\frac{C}{n^2} - \frac{C}{m^2} = h \cdot \nu$$

$$\nu = \frac{C}{h}\left(\frac{1}{n^2} - \frac{1}{m^2}\right).$$

Hier könnte man nun sagen, es sei ganz deutlich eine Zielstrebigkeit des Elektrons zu erkennen: das Elektron muß doch im voraus wissen, wohin es springt, sonst kann es ja gar nicht die richtige Strahlung aussenden. Dennoch denkt heute kein Physiker daran, dem Elektron seelische Eigenschaften zuzuschreiben, sondern man sagt sich, daß die Bohrsche Regel auf tiefer liegende Gesetze zurückgeführt werden muß, welche wir nur gegenwärtig noch nicht kennen. In diesem Fall ist uns also nur das Integralgesetz bekannt; das Differentialgesetz dagegen müssen wir erst noch suchen. Wir würden es aber für eine müßige Ausflucht halten, wenn jemand das Aufsuchen eines solches Gesetzes durch die Annahme von seelischen Eigenschaften der Elektronen ersetzen wollte.

9. KAPITEL: DIE ROLLE VON IDEALEN GEBILDEN

Zum Schlusse möchte ich noch von einem Thema handeln, das dem Mathematiker besonders nahe steht: *von den idealen Gebilden.* Das Ideale wird dem Realen gegenübergestellt, und zwar einerseits als das Unwirkliche (bloß Gedachte) dem Wirklichen, andererseits als das Vollkommene dem existierenden Unvollkommenen. Ich will zunächst nur von der ersten Bedeutung des Idealen sprechen. Diese spielt besonders in der Mathematik eine große Rolle. Auf den ersten Blick erscheint diese Tatsache sehr paradox; denn da ja die Mathematik es nicht unmittelbar mit dem Wirklichen zu tun hat, sondern sich im Reiche des reinen Gedankens bewegt, so versteht man nicht, wie innerhalb der Mathematik eine Unterscheidung zwischen Wirklichem und Unwirklichem stattfinden kann.

Demgegenüber ist nun folgendes zu bemerken: wenn auch die Mathematik nicht von dem Wirklichen im gewöhnlichen Sinne handelt, so ist doch sehr viel von Existenz in der Mathematik die Rede. Gehören doch die Existenzbeweise gerade zu den wichtigsten und schwierigsten mathematischen Problemen. Was ist aber hier mit Existenz gemeint? Sieht man näher zu, so findet man, daß von Existenz immer mit Bezug auf ein bestimmtes *zugrunde gelegtes System* gesprochen wird, und zwar ist dieses System ein anderes je nach der Theorie, welche man behandelt. So hat man es in der elementaren Geometrie mit dem System der Punkte, Geraden und Ebenen zu tun, welches durch die Axiome charakterisiert wird, in der Zahlentheorie mit dem System der ganzen Zahlen, welche man, von der Eins ausgehend, durch sukzessive Hinzufügung von Eins erhält.

Hiernach läßt sich nun leicht erklären, inwiefern in der Mathematik ideale Gebilde vorkommen. Dies findet nämlich statt, wenn ein System durch Einführung neuer Elemente erweitert wird, die dann für das ursprüngliche System den Charakter idealer Elemente haben.

Ein gutes Beispiel hierfür bildet die Frage der Wurzelexistenz in der Algebra. Zieht man hier als Lösungen nur reelle Zahlen in Betracht, so muß man von manchen Gleichungen sagen, daß sie keine Wurzeln haben. Hierbei könnte man es auch bewenden lassen. Doch würde dann die allgemeine Theorie der Gleichungen zu schwerfällig sein. Man erweitert deshalb das System der reellen Zahlen durch Einführung der komplexen Zahlen, wodurch die ausnahmslose Wurzelexistenz erreicht wird.

Dieses Verfahren der Einführung idealer Elemente ist eine der wichtigsten mathematischen Methoden, deren Anwendung sich bis in die höchsten Teile der Mathematik immer wiederholt. Genau genommen besteht das Verfahren darin, daß man von einem ursprünglichen System, welches sich in Hinsicht auf gewisse zu behandelnde Fragen kompliziert verhält, zu einem neuen System übergeht, bei welchem sich jene Verhältnisse einfacher gestalten und welches außerdem die Eigenschaft hat, ein Teilsystem zu enthalten, daß mit dem ursprünglichen System isomorph ist (d. h. mit diesem in allen darin vorliegenden Beziehungen übereinstimmt). Es ergeben sich dann die Sätze, welche sich auf das ursprüngliche System beziehen, aus den Sätzen für das neue System als spezielle Folgerungen, indem man die Bedingungen berücksichtigt, welche das betreffende Teilsystem charakterisieren. Und der Vorteil dabei ist, daß die Ausführung der Beweise nun innerhalb des neuen Systems geschieht, wo alles viel übersichtlicher wird.

Die Redeweise von den idealen Elementen hat hierbei eigentlich nur vom Standpunkt der Betrachtung des Ausgangssystems ihre Berechtigung. In dem neuen System unterscheiden wir gar nicht mehr zwischen wirklichen und idealen Elementen. Der Übergang zu dem weiteren System kann entweder auf konstruktivem Wege geschehen, indem die neuen Elemente durch mathematische Konstruktion aus den alten abgeleitet werden, oder auf axiomatischem Wege, indem das neue System durch Relationseigenschaften charakterisiert wird. In dem zweiten Falle bedarf es noch des Nachweises, daß die Annahme eines Systems von der verlangten Beschaffenheit keinen Widerspruch in sich schließt.

Von der beschriebenen Methode macht man bereits in der *elementaren Arithmetik* Gebrauch. Hier werden zunächst nur die positiven ganzen Zahlen als existierende Zahlen betrachtet. Dann aber erweitert man das Zahlensystem durch die Einführung der negativen und der gebrochenen Zahlen. Der Zweck ist hierbei ganz deutlich. Man verschafft sich auf diese Weise den Vorteil, daß die elementaren Rechenoperationen unbeschränkt ausführbar werden (ausgenommen nur die Division mit Null). Das System der rationalen Zahlen, welches man durch Hinzunahme der neuen (idealen) Elemente zu dem System der positiven ganzen Zahlen erhält, besitzt in der Tat wesentlich einfachere Gesetzmäßigkeiten als jenes ursprüngliche System. Zum Beispiel gelten hier die einfachen Sätze der Determinantentheorie über die Auflösung linearer Gleichungen. Die Erweiterung des Systems der positiven ganzen zu dem der rationalen Zahlen läßt sich konstruktiv durchführen, indem man die Brüche als Paare von positiven ganzen Zahlen (unter geeigneter Definition der Gleichheit) und entsprechend die negativen rationalen Zahlen als Paare von Brüchen erklärt.

Von den weiteren sehr mannigfachen Anwendungen, welche die Methode der idealen Elemente in der Mathematik erfährt, will ich noch zwei etwas näher besprechen. Die eine findet bei der *Euklidischen Geometrie* statt. Beschränken wir uns hier auf die Geometrie der Ebene, so haben wir als existierende Gegenstände die Punkte und Geraden in der Ebene zu betrachten. Für diese gilt das Axiom der Verknüpfung: Durch zwei Punkte geht immer eine und nur eine Gerade. Und hieraus ergibt sich als Folgerung, daß zwei Geraden sich höchstens in einem Punkte schneiden. Es gilt jedoch nicht der Satz, daß zwei Geraden sich immer in einem Punkte schneiden. Wir können nun aber durch Einführung idealer Schnittpunkte und Geraden erreichen, daß dieser Satz sich in allgemeiner Form aufstellen läßt. Das Verfahren der Erweiterung gestaltet sich hierbei deshalb besonders einfach, weil wir in der Euklidischen Geometrie die einfachen Sätze über die Parallelen haben, wonach zu einer gegeben Geraden durch einen außerhalb gelegenen Punkt immer eine und nur eine Gerade hindurchgeht, welche die erste Gerade nicht schneidet. Aufgrund hiervon genügt es, um die gewünschte Vereinfachung zu erhalten, wenn wir folgende Festsetzungen treffen:

1) Jede Gerade enthält einen idealen Punkt.

2) Je zwei parallele Geraden haben denselben idealen Punkt.

3) Zwei sich schneidende Geraden haben verschiedene ideale Punkte.

4) Alle idealen Punkte liegen auf einer idealen Geraden, (welche also die einzige ideale Gerade ist).

Diese Festsetzungen sind notwendig und hinreichend, damit man, nach Einbeziehung der idealen Elemente, sagen kann: Zwei Punkte liegen immer auf einer und nur einer Geraden, und zwei Geraden haben immer einen und nur einen Punkt gemeinsam.

Zugleich mit dieser wesentlichen Vereinfachung stellt sich hier noch ein weiterer Vorteil ein, welchen man gewahr wird, wenn man für die Theorie des erweiterten Systems, die "projektive Geometrie", die Axiome aufstellt, wie sie sich durch sinngemäße Modifikation der ursprünglichen Axiome ergeben. Diese Axiome haben nämlich die Eigenschaft, daß sie in sich übergehen, wenn man in ihrer Formulierung die Worte "Punkt" und "Gerade" vertauscht. Auf diesem Umstand gründet sich das Dualitätsprinzip der projektiven Geometrie, welches besagt, daß man aus jedem Satz der projektiven Geometrie (der Ebene) wieder einen richtigen Satz erhält, indem man Punkt und Gerade ihre Rollen vertauschen läßt.

Als ein Beispiel für die Anwendung dieses sehr fruchtbaren Prinzips will ich folgenden Satz nehmen, welcher einen Spezialfall des berühmten Pascalschen Satzes bildet: wenn die Ecken eines (überschlagenen) Sechsecks abwechselnd auf zwei Geraden liegen, so liegen die Schnittpunkte der gegenüberliegenden Seiten auf einer Geraden.

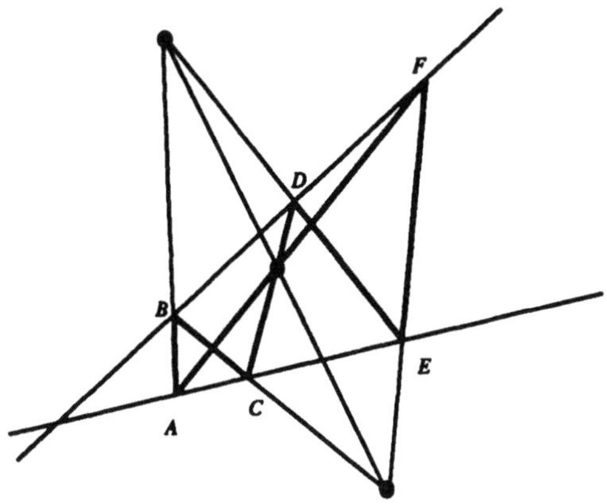

Fig. 7

Gemäß dem Prinzip der Dualität folgt aus diesem Satz der zu ihm duale: Wenn die Seiten eines Sechsseits abwechselnd durch zwei Punkte gehen, so schneiden sich die Verbindungslinien der gegenüberliegenden Ecken in einem Punkt. Diesen Satz wiederum können wir für die gewöhnliche Euklidische Geometrie auch so spezialisieren, daß wir die zwei Punkte als ideale Punkte wählen; er nimmt dann folgende Form an: Wenn in einem Sechseck die geradzahligen sowie die ungeradzahligen Seiten einander parallel sind, so schneiden sich die Verbindungslinien der gegenüberliegenden Ecken in einem Punkt.

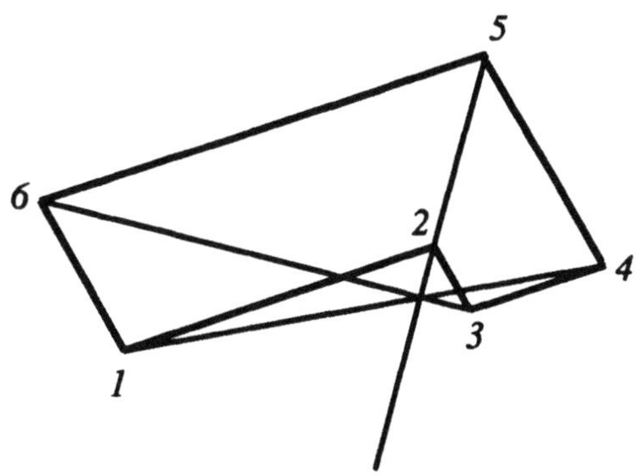

Fig. 8

Hier zeigt sich deutlich, was wir durch die Methode der projektiven Geometrie gewinnen. Sie ermöglicht es uns, verschiedene Sätze, die von dem ursprünglichen Standpunkt als ganz getrennt erscheinen, auf einmal zu beweisen, indem sie nunmehr bloß als verschiedene Wendungen und Spezialisierungen eines und desselben Satzes aufgefaßt werden.

Der Nachweis der Widerspruchslosigkeit der projektiven Geometrie der Ebene kann durch die Zurückführung auf die Geometrie des Strahlenbündels (im gewöhnlichen Euklidischen Raum) erbracht werden. Man projiziert hierzu die Ebene von einem außerhalb gelegenen Punkte und läßt den Punkten der Ebene die hindurchgehenden Projektionsstrahlen, den Geraden der Ebene die durch sie gehenden Projektionsebenen entsprechen. Den zu der Ebene parallelen Strahlen des Bündels ordnet man die idealen Punkte zu, und der durch den Strahlenmittelpunkt gehenden Parallel-Ebene die ideale Gerade. (Eine andere Möglichkeit ist, die projektive Geometrie auf der Kugeloberfläche zu deuten, wobei je ein Paar von gegenüberliegenden Punkten als ein einziger Punkt anzusehen ist und die größten Kreise die Rolle der Geraden spielen.)

Die Geometrie der Ebene liefert uns auch ein gutes Beispiel dafür, daß bei einem vorliegenden Ausgangssystem *nicht nur eine* Möglichkeit der Erweiterung besteht, daß vielmehr die Art, wie man das System erweitert, je nach dem Zweck, welchen man verfolgt, eine andere sein kann. Wir finden nämlich in der komplexen Funktionentheorie eine andere Erweiterung der elementaren Geometrie der Ebene. Hier wird zu den Punkten der Ebene, welche durch die komplexen Zahlen dargestellt sind (die Zahl $u + iv$ stellt den Punkt mit den Koordinaten u, v dar), nur ein einziger idealer Punkt hinzugefügt. Dies tut man in der Theorie der rationalen Funktionen, um allgemein den Satz aussprechen zu können, daß eine rationale Funktion jeden komplexen Wert gleich oft annimmt. Insbesondere ist die Einführung des idealen Punktes bei der Untersuchung der komplexen linearen Transformationen von Nutzen. Hier spielen die Kreise zusammen mit Geraden eine ausgezeichnete Rolle, und die Geraden werden aufgefaßt als diejenigen Kreise, welche durch den idealen Punkt gehen.

Das zweite typische Beispiel einer Einführung idealer Elemente, das ich noch besprechen will, ist die *Theorie der idealen Teiler* in der Zahlentheorie. Den ersten Anstoß zu dieser Theorie hat die Bemerkung gegeben, daß der Satz von der eindeutigen Zerlegung der ganzen (rationalen) Zahlen in Primfaktoren sich auch auf andere Zahlbereiche ausdehnen läßt. Zum Beispiel gilt der Satz auch für den Bereich der Zahlen $a + bi$, wo a, b ganze rationale Zahlen bedeuten. Unter einer Primzahl versteht man hier eine solche Zahl, die sich nicht weiter in Faktoren zerlegen läßt. Dabei wird die Abspaltung einer der Zahlen $+1$, -1, $+i$, $-i$ nicht als Zerlegung gerechnet; diese Zahlen spielen hier die Rolle der Einheiten, denn das Reziproke einer jeden von ihnen ist wieder eine Zahl des Bereiches.

Die Zahl 3 ist auch in diesem Bereich eine Primzahl; denn gäbe es eine Zerlegung

$$3 = (a + bi)(c + di),$$

so müßte auch

$$3 = (a - bi)(c - di),$$

folglich

$$9 = (a^2 + b^2)(c^2 + d^2)$$

sein, was nur möglich ist, wenn entweder

$$a^2 + b^2 = 1$$

oder

$$c^2 + d^2 = 1,$$

also eine der Zahlen $a+bi$, $c+di$ eine Einheit ist, entgegen unserer Annahme.

Dagegen ist 5 keine Primzahl; denn es besteht die Zerlegung

$$5 = (1 + 2i)(1 - 2i).$$

Ebensowenig ist 2 eine Primzahl; denn

$$2 = -i(1 + i)^2.$$

(Von den hier als Faktoren auftretenden Zahlen $1+2i$, $1-2i$, $1+i$ zeigt man leicht, daß sie Primzahlen sind.)

Als Beispiel einer etwas komplizierten Zerlegung in Primfaktoren will ich anführen:

$$-306 + 1530i = -1(1 - i)^3 3^2 (4 + i)(4 - i)(2 + 3i).$$

Nun gelingt aber die Übertragung der gewöhnlichen Zahlentheorie nicht bei allen Bereichen dieser Art. Betrachten wir z. B. die Zahlen von der Form $a + b\sqrt{-5}$, (wo wieder a, b ganze rationale Zahlen sein sollen). Unter diesen haben nur $+1$ und -1 die Eigenschaft der Einheiten. Definiert man hier nun wieder als Primzahlen die unzerlegbaren Zahlen, so ist die Zerlegung in Primfaktoren nicht eindeutig. Bereits für die Zahl 6 gibt es zwei verschiedene solche Zerlegungen:

$$6 = 2 \cdot 3$$

und

$$6 = (1 + \sqrt{-5})(1 - \sqrt{-5}).$$

(Daß die Zahlen 2, 3, $1 + \sqrt{-5}$, $1 - \sqrt{-5}$ unzerlegbar sind, ist leicht zu beweisen. Gäbe es z. B. für 2 eine Zerlegung

$$2 = (a + b\sqrt{-5})(c + d\sqrt{-5}),$$

so würde folgen

$$2 = (a - b\sqrt{-5})(c - d\sqrt{-5}),$$

$$4 = (a^2 + 5b^2)(c^2 + 5d^2),$$

und diese Gleichung könnte nur bestehen, wenn eine der Zahlen $a+b\sqrt{-5}$, $c+d\sqrt{-5}$ gleich 1 wäre.)

Nun könnte man sich mit der Feststellung dieser Tatsache begnügen und auf die Eindeutigkeit der Zerlegung in Primfaktoren bei diesem and anderen solchen Zahlbereichen verzichten. Dadurch würde aber bei den höheren zahlentheoretischen Untersuchungen, z. B. in der Theorie der Kreisteilung, alles sehr unübersichtlich werden. Es war daher eine sehr wichtige Entdekkung, als Kummer bemerkte, daß die Eindeutigkeit der Primfaktorenzerlegung für die (endlichen) algebraischen Zahlkörper sich durch die Einführung idealer Primteiler aufrecht erhalten läßt.

Auf den Gedanken, ideale Primteiler einzuführen, könnte man durch folgende Analogiebetrachtung aus dem Gebiet der ganzen rationalen Zahlen gebracht werden. Angenommen, es wären uns aus irgendeinem Grunde nur diejenigen Zahlen bekannt, welche, im Dezimalsystem geschrieben, auf 1 oder 6 endigen, also

$$1, 6, 11, 16, 21, 26, \ldots.$$

Dann würde die Zerlegung der Zahlen in unzerlegbare Faktoren nicht eindeutig sein, wie folgende Beispiele beweisen

$$336 = 6 \cdot 56 = 16 \cdot 21$$

$$1296 = 6^4 = 16 \cdot 81,$$

wo 6, 16, 56, 21, 81 im Gebiete der als bekannt angenommenen Zahlen unzerlegbar sind. Um hier nun die Eindeutigkeit der Zerlegung zu erhalten, müßte man andere Zahlen hinzunehmen, und zwar würden diese als ideale größte gemeinsame Teiler von bekannten Zahlen einzuführen sein, (z. B. die Zahl 3 als größter gemeinsamer Teiler von 6 und 21).

In entsprechender Weise verfährt man auch bei den Zahlbereichen der algebraischen Zahlkörpertheorie. Dabei ist es besonders beachtenswert, daß die Einführung der idealen Teiler auf rein konstruktivem Wege, nach einer von Dedekind angegebenen Methode, gelingt. Man definiert als größten gemeinsamen Teiler zweier Zahlen α, β den Inbegriff aller Zahlen von der Form

$$u \cdot \alpha + v \cdot \beta,$$

wo u, v Zahlen des betrachteten Bereiches sind, und wendet dafür das Symbol (α, β) an. Entsprechend definiert man den größten gemeinsamen Teiler von mehreren Zahlen, so daß z. B. $(\alpha, \beta, \gamma, \delta)$ den Inbegriff der Zahlen von der Form

$$u \cdot \alpha + v \cdot \beta + w \cdot \gamma + t \cdot \delta$$

bedeutet, wo wieder u, v, w, t Zahlen des Bereiches sein sollen. Unter dem Produkt zweier solcher idealer Teiler versteht man den Inbegriff aller der Zahlen, die sich als Produkt aus einer zu dem ersten Teiler und einer zu dem zweiten Teiler gehörigen Zahl darstellen lassen. Dieses Produkt kann dann wiederum als größter gemeinsamer Teiler von endlich vielen Zahlen des Bereiches dargestellt werden. Zum Beispiel ist das Produkt von (α, β) und (γ, δ) identisch mit dem idealen Teiler

$$(\alpha \cdot \gamma, \alpha \cdot \delta, \beta \cdot \gamma, \beta \cdot \delta).$$

(Die Darstellung eines idealen Teilers als größter gemeinsamer Teiler von Zahlen ist natürlich auf sehr verschiedene Weisen möglich; z. B. ist $(\alpha, \alpha + \beta, \alpha - \beta)$ derselbe Teiler wie (α, β).)

Aufgrund der getroffenen Festsetzungen geschieht nun die Zerlegung einer Zahl α in der Weise, daß man zunächst anstelle der Zahl α den Inbegriff der durch α teilbaren Zahlen des Bereiches betrachtet. Dieser bildet einen speziellen idealen Teiler, und man zerlegt nun diesen soweit in ideale Faktoren, daß diese ihrerseits nicht mehr als Produkte von idealen Teilern darstellbar sind. (Dabei ist der von sämtlichen Zahlen des Bereiches gebildete ideale Teiler, welchem die Rolle einer Einheit zukommt, nicht als Faktor zu rechnen.)

Bei diesem Verfahren wird in der Tat das Ziel der Eindeutigkeit der Zerlegung erreicht. Man kann beweisen, daß die Zerlegung einer Zahl in ideale Primteiler (d. h. in nicht weiter zerlegbare ideale Teiler) nur auf eine einzige Art möglich ist. So besteht z. B. in dem durch ideale Teiler erweiterten Bereich der Zahlen $a + b\sqrt{-5}$ für die Zahl 6 die Zerlegung

$$(2, 1 + \sqrt{-5})^2 (3, 1 + \sqrt{-5})(3, 1 - \sqrt{-5}),$$

wo $(2, 1 + \sqrt{-5})$, $(3, 1 + \sqrt{-5})$, $(3, 1 - \sqrt{-5})$ ideale Primteiler sind; und eine andere solche Zerlegung von 6 gibt es nicht.

Durch die Einführung von idealen Teilern sind wir also imstande, die Zerlegungsgesetze der gewöhnlichen Zahlentheorie bei den algebraischen Zahlkörpern allgemein aufrecht zu erhalten. Hierauf beruht es, daß die idealen Teiler eine wesentliche Rolle in der höheren Zahlentheorie spielen. So ist es z. B. erst durch die Heranziehung der idealen Teiler gelungen, die Unmöglichkeit einer Lösung der Fermatschen Gleichung

$$x^p + y^p = z^p$$

durch ganze rationale Zahlen x, y, z für alle ungeraden Primzahlen p unter 100 zu beweisen, während man vordem nur in den Fällen $p = 3, 5, 7$, und auch hier nur durch sehr mühsame Überlegungen, den Beweis führen konnte.

Die Methode der Einführung idealer Elemente, welche ich an diesen verschiedenen Beispielen dargelegt habe, ist eine spezifisch mathematische. Dennoch finden wir ein analoges Verfahren auch auf anderen Gebieten des Denkens. Ein gutes Beispiel bildet in der Rechtslehre der Begriff der *juristischen Person*, die als Trägerin von Rechten zu den wirklichen Personen hinzugedacht wird.

Von einem gewissen philosophischen Standpunkt, welcher nur die Inhalte der Wahrnehmung als etwas Reales gelten läßt, werden überhaupt alle Gegenstände der objektiven physikalischen Naturbetrachtung bloß als ideale Gebilde aufgefaßt, die wir einführen, um unsere Wahrnehmungserlebnisse übersichtlich ordnen zu können. Insbesondere ist nach dieser Auffassung alles das als idealer Gegenstand anzusehen, was wir nur nach Analogie der aus der Wahrnehmung bekannten Dinge annehmen, wie z. B. die Atome und Elektronen. Das gesamte physikalische Denken und das analoge Denken im Täglichen erscheint für diese Ansicht als nichts anderes als eine unausgesetzte Anwendung der betrachteten Methode der Erweiterung eines Systems durch Einführung von idealen Elementen.

Nun gibt es aber noch eine ganz *andere Art von Idealisierung*, die für die reine Mathematik nicht in Betracht kommt, in der Physik aber reiche Anwendung findet. Bei der mathematischen Idealisierung ist das Wesentliche der Übergang zu einem größeren System, bei welchem die Beziehungen übersichtlicher werden. Die andere Methode des Idealisierens dient ebenfalls dem Zweck der Vereinfachung, und zwar wird diese hier dadurch bewirkt, daß man von gewissen Komplikationen absieht. Man betrachtet einen gegebenen Gegenstand als *Annäherung an ein Ideal*, das durch größere Einfachheit ausgezeichnet ist. Hier tritt uns die zweite Bedeutung des Idealen entgegen: der wirkliche Gegenstand erscheint als eine nur unvollkommene Realisierung eines begrifflich bestimmten Vorbildes. Das ideale Gas, die ideale Flüssigkeit, der vollkommen elastische Körper, der freie Fall, die Zwangsbedingungen in der Mechanik sind bekannte Beispiele solcher Idealisierungen.

Allgemein ist zu sagen: wenn wir einen Gegenstand der Wirklichkeit physikalisch beschreiben, so besteht das darin, daß wir ihn sukzessive durch ideale Gegenstände, d. h. durch Begriffsgebilde approximieren. Diese fortschreitende Annäherung betrifft insbesondere auch die Bestimmung der Größenverhältnisse. Eine endgültige Bestimmung einer physikalischen Größe ist (wegen der Ungenauigkeit aller Messungen) nur da denkbar, wo wir es mit Diskretem zu tun haben, wo also das Messen auf ein Zählen hinauskommt. Dies ist auch einer der Gründe für die Tendenz der Physik zur Atomistik. In einer rein atomistischen Theorie wäre eine vollkommen genaue Bestimmung der physikalischen Konstanten wenigstens grundsätzlich möglich. Zugleich damit würde man den Vorteil haben, daß die Frage der Maßeinheiten sich hier durch das Vorhandensein letzter Einheiten ganz von selbst erledigte. Eine *natürliche Wahl der Maßeinheiten* läßt sich übrigens auch vom Standpunkt der Feldtheorie treffen.

Es ist Planck gelungen, zu zeigen, daß man die Willkür der üblichen Maßeinheiten (Meter, Sekunde usw.) beseitigen kann, indem man alles durch die Lichtgeschwindigkeit, die Gravitationskonstante, die Boltzmannsche Konstante k und das elementare Wirkungsquantum h ausdrückt. Der wesentliche Gedanke hierbei ist, daß man nicht wie bisher für die Festsetzung der Maßeinheiten spezielle Integrale der Differentialgleichungen (z.B die Länge des Erdradius, die Dauer einer bestimmten Pendelschwingung) benutzt, sondern die *Konstanten in den Differentialgleichungen* festlegt. Bei der Einführung eines solchen Maßsystems müssen alle physikalischen Konstanten reine Zahlen werden.

Diese Ausschaltung der Willkür in den Maßeinheiten bedeutet auch einen Fortschritt in Einsicht auf das Ziel der Idealisierung, indem dadurch die Beschreibung der Größenverhältnisse den Begriffsbildungen der Theorie so eng wie möglich angepasst wird. Wenn man nun die immer weiter gehende Anpassung der physikalischen Begriffsbildungen an die Wirklichkeit betrachtet, so wird der Gedanke an eine *Grenze* dieses Prozesses nahegelegt. Das hierbei vorschwebende Endziel besteht in einer vollständigen mathematischen Beschreibung der gesamten Wirklichkeit, wie ich sie zu Beginn unserer Erörterungen über die physikalische Gesetzlichkeit in Betracht gezogen habe.

Wir wollen uns nun klar machen, wie dieses *Limesideal* beschaffen sein müßte, indem wir uns dabei an die moderne Feldtheorie halten, durch welche das physikalische Limesideal eine bestimmtere Form gewinnt. Im Sinne der grundsätzlichen Auffassung dieser Theorie, wie sie in meinen Noten formuliert ist, haben wir anzunehmen, daß das Naturgeschehen vollständig bestimmt ist durch den Wertverlauf von 14 Funktionen der Koordinaten x, y, z, t, nämlich den 10 Gravitationspotentialen $g_{\mu\nu}$, welche die Komponenten eines symmetrischen Tensors bilden, und den 4 elektrischen Potentialen q_ν, welche die Komponenten eines Vektors sind. Hiernach würde die Vollendung der Physik in der vollständigen Kenntnis dieser 14 Funktionen bestehen. Gemäß unseren früheren Überlegungen wäre eine solche vollkommene Kenntnis in der Weise zu denken, daß die Funktionen durch ein System von Differentialgleichungen und gewisse hinzutretende Nebenbedingungen (bzw. Randbedingungen) eindeutig festgelegt würden (abgesehen von der durch die Koordinatenwahl bedingte Willkür).

Durch die Gedanken der neuesten *Weylschen Theorie* wird nun noch die Möglichkeit gegeben, das System der Funktionen $g_{\mu\nu}$ und q_ν einheitlich als eine Maßbestimmung der 4-dimensionalen Raum-Zeit-Mannigfaltigkeit zu deuten. Wir kommen damit zu der Konsequenz, daß der physikalische Zustand an einer Raum-Zeitstelle vollkommen beschrieben wird durch das an der Stelle geltende Raum-Zeit-Maß. Das gesamte physikalische Geschehen würde sich nach dieser Auffassung darstellen als eine bestimmte räumlich-zeitliche Geometrie, und das Ziel der Physik würde erreicht sein mit der vollständigen Kenntnis dieser Maß-Geometrie des Raum-Zeit-Kontinuums. In diesem physikalischen Weltbild haben wir ein Beispiel einer extremen Idealisierung. Die gesamte Natur wird sozusagen einem einzigen idealen Gegenstande gleichgesetzt, dessen Eigenschaften durch die Physik beschrieben werden, entsprechend wie nach der alten Auffassung die Geometrie die Eigenschaften des Raumes beschreibt.

Die Sonderstellung der Lehre vom Raum ist bei dem Limesideal der modernen Physik ganz aufgehoben. Wir haben da nicht eine vorphysikalische abgesonderte Raum-Zeit-Theorie, auf welche sich erst die Physik aufbaut, sondern eine einheitliche physikalische Theorie, aus welcher sich die geometrischen und kinematischen Beziehungen als Folgerungen ergeben. Vorweggenommen wird nur die topologische Beschaffenheit des Raum-Zeit-Kontinuums sowie der indefinite Charakter des Linienelementes. Die Methode der Differentialgeometrie, mit Hilfe deren man ursprünglich die Eigenschaften des Raumes (bzw. der Flächen im Raume) studierte, wird nun die Methode für die mathematische Theorie der äußeren Natur überhaupt. Man betrachtet also die ganze Physik, nach Analogie der Lehre vom Raum, als eine allgemeine Geometrie.

Wenn wir nun aber dieses Weltbild näher ins Auge fassen, welches sich aus der Vorstellung des vollendeten physikalischen Idealisierungprozesses ergibt, so werden wir gewahr, daß wir damit in *Paradoxien* geraten. Eine erste solche Paradoxie besteht darin, daß bei der vollständigen physikalischen Bestimmtheit der Erfüllung von Raum und Zeit auch das *organische Geschehen* mit inbegriffen sein würde, und daß daher die Vorgänge der belebten Natur den gleichen Charakter der *zeitlichen Umkehrbarkeit* besitzen müßten wie die anorganischen Vorgänge. Es könnte danach also auch für das organische Geschehen keine Auszeichnung einer Zeitrichtung geben.

Eine weitere Paradoxie hängt damit zusammen, daß in der alles umfassenden physikalischen Theorie dasjenige vorliegen würde, was ich eine *Hegelsche Physik* genannt habe. Es würde da alles, was noch geschehen wird, endgültig vorbestimmt sein. Eigentliche Entscheidungen könnten gar nicht stattfinden, und das ganze Weltgeschehen würde nicht über den *begrenzten Inhalt eines endlichen Gedankens* hinausgehen.

Vor allem aber besteht die Schwierigkeit, daß unter der Annahme einer vollständigen mathematischen Naturbeschreibung die Wirklichkeit mit einem idealen Gegenstande, d. h. einem wesenlosen Gedankendinge identisch sein müßte. Denn es bliebe ja gar nichts Qualitatives, was noch hinzukommen könnte, da ja alle *Qualitäten* auf mathematische Beziehungen zurückgeführt und somit als *bloß scheinbar* anzusehen wären. Aus demselben Grunde müßte auch das *Geistige*, und insbesondere unser Denken, etwas bloß Scheinbares sein—eine absurde Konsequenz für eine Naturansicht, welche hervorgeht aus dem Bestreben, alle Inhalte der Wirklichkeit unserem Denken zugänglich zu machen.

Durch diese Paradoxien werden wir zu der Folgerung genötigt, daß die Annahme der Vollendbarkeit des physikalischen Idealisierungprozesses unzulässig ist, daß also das physikalische Limesideal grundsätzlich unerreichbar ist. Diese Tatsache braucht uns nicht sehr zu verwundern, da wir uns ja in der Mathematik daran gewöhnt haben, die Erreichung eines Grenzwertes als etwas keineswegs Selbstverständliches zu betrachten. Wie diese Unvollendbarkeit zu verstehen ist, kann erst in tiefstem Sinne klar werden, wenn man zuvor die mengentheoretische Paradoxie von der Unmöglichkeit der Beschreibung aller Gesetze durch eine Anzahl von Worten gelöst hat.

Bemerkenswert ist aber, daß wir bei der Erörterung der Methode der Physik durch die vorgefundenen Paradoxien auf das allgemeine philosophische Problem geführt werden, ob und wie es möglich ist, unser Denken durch das Denken selbst zu begreifen und es von jeglichen Paradoxien zu befreien. Es ist dies dieselbe Frage, welche auch unseren Bemühungen im Gebiete der mathematischen Logik zugrunde liegt.

MIX
Papier aus verantwortungsvollen Quellen
Paper from responsible sources
FSC® C105338

If you have any concerns about our products,
you can contact us on
ProductSafety@springernature.com

In case Publisher is established outside the EU,
the EU authorized representative is:
Springer Nature Customer Service Center GmbH
Europaplatz 3, 69115 Heidelberg, Germany

Printed by Libri Plureos GmbH
in Hamburg, Germany